Our Universe Rotates inside God's Mighty Hand
(The Speed of God)

Ifeanyi Chukwujama

This Book is available at:
www.amazon.com
www.kindle.com
www.createspace.com
www.Jesus-On.com

For permission to reprint or copy this booklet, please contact the publisher:

Ifeanyi Chukwujama

Jesus-On Kingdom Ministries

394-396 Warren Street, Boston, MA 02119

Our email: JesusOnkm@gmail.com

Our website: www.Jesus-On.com

ISBN-10: 1507784392
ISBN-13: 978-1507784396

DEDICATION

To my children and my wife! I thank God Almighty for allowing me to live long enough to teach my children that the most important possession in this life is having an intimate relationship with God and His Son Jesus Christ through obedience and complete surrender of our lives to Jesus Christ!

Also to my readers who come to share the same value in Jesus Christ! May our merciful and loving God maintain His perpetual love in our hearts, and keep us under his glorious wings forever and ever. Amen!

CONTENTS

ACKNOWLEDGMENTS

The only teacher any of us needs in this world is God. His word imprints on your heart as a tattoo and lives inside of you throughout your life. We are living in the end times when God promises to write His laws on the hearts of men that no one need to say to his brother, "Do you know the Lord?" Because we will all know God and revel in his love and mercy. Desire the Lord with all your heart, and all your soul and all your mind and His grace will abound in your life.

The Bible is the source of all knowledge. Embrace it and live

Chapter 1
Our Universe fits into God's Mighty Hand; and rotates

The mysteries of God are unending! Rejecting God puts one in perpetual blindness. This time, it is the world of science that is shown to be in pitch darkness for its ignorance of God. That is why, speaking of the eyes of our spirit, the Bible says:

""The eye is the lamp of the body. If your eyes are healthy, your whole body will be full of light. But if your eyes are unhealthy, your whole body will be full of darkness. If then the light within you is darkness, how great is that darkness!" (Matthew 6:22-23)

God has revealed that in their desperate attempt to remove God's name from everything in the world, scientists have missed out on the fastest speed the universe has ever experienced—the **"speed of God,"** which is infinitely faster than the "speed of light" at 186,000 miles per second.

God created the earth and the universe and everything in them at the "speed of God"; not at the "speed of light". By looking at everything in the world and the universe through the speed of light, science has misled the world into believing in billions of years of creation. This is only a science fiction; and not science!

The truth: All of creation—the earth and the universe and everything in them—took God only six days to complete; working at the "speed of God."

Listen to God summarize His creation work:

3

"It is I who made the earth and created mankind on it. My own hands stretched out the heavens; I marshaled their starry hosts." (Isaiah 45:12)

"Lift up your eyes and look to the heavens: Who created all these? <u>He who brings out the starry host one by one and calls forth each of them by name.</u> Because of his great power and mighty strength, not one of them is missing." (Isaiah 40:26)

"My own hand laid the foundations of the earth, and my right hand spread out the heavens; **<u>when I summon them, they all stand up together.</u>**" (Isaiah 48:13)

And to those who take pride in snubbing God and mischaracterizing His work and His great intentions for mankind, God says: *"All day long I have held out my hands to an obstinate people, who walk in ways not good, pursuing their own imaginations"* (Isaiah 65:2)

For instance, scientists recently identified the distant Galaxy MACS0647-JD and "based on a photometric redshift estimate", dubbed it the farthest known galaxy from the Earth, claiming that it formed 13.3 billion years ago. This claim is a great error of science because the Bible—which is more science than any other created thing is the universe—says that the entire universe came into place within 24 hours of Day 4 of creation.

The earth, which science has estimated to be 4.5 billion years old—as compared to 13.8 Billion years projection for the universe—is, in reality, the oldest thing in the entire universe *(Genesis 1:2)*. **The earth is older than time, space and the universe itself!**

This is God's undisputed truth. So, who knows it best? God or overzealous scientists who tickle their own fancy by picking a fight with God? Anyone who believes creation by evolutionary process is yet to embrace the truth, because trusting the word of God is the only way anyone can come to the truth.

The scientists based their calculations on the speed of light because they assumed that the universe created itself; and that from the moment of the "Big Bang", everything travelled away from the earth to their current locations, at the speed of light—since light is the fastest thing in the universe. But their estimation is in great error. And their error was caused by their deliberate decision to leave God out of creation.

God is the substance of creation! And leaving the substance out of anything leaves one with emptiness. When you choose the wrong reference point, all your measurements and computations will all come out wrong. God is the only reference point to everything that exists on the earth and in the universe.

God did not create the earth and the universe at the speed of light. God created everything at the "speed of God", which is infinitely faster than the speed of light. Thus, the 13.3 billion years origination estimate for the Galaxy MACS0647-JD, in reality, took God only a few seconds; since all the other billions of galaxies in the universe were also placed in location and set in clusters and motions within the 24 hours of Day 4 of creation.

The speed of God created the universe; and not the speed of light. Scientists' great error by throwing away the "speed of God" has also misled them into mischaracterizing the true mode of propagation of the light they see from distant galaxies, because one erroneous assumption leads to another.

Working at the speed of God, God created space in the 24 hours that constituted Day 2 of creation *(Genesis 1:6-8)*. Then in another 24 hours that constituted Day 4 of creation *(Genesis 1:14-19)*, God populated space completely, to form our universe. This is why God said in the Book of Isaiah: *"My own hand laid the foundations of the earth, and my right hand spread out the heavens; **when I summon them, they all stand up together**" (Isaiah 48:13)*. The last sentence alludes to God's commands in Genesis Chapter One

5

which brought each of the creations into existence.

The Bible says that our entire Universe fits into God's mighty hand *(Isaiah 40:12)*; and rotates in God's hand *(2 Peter 3:8)*, suspended over nothing *(Job 26:7)*. The Bible also says that our earth is right in the center of the universe *(Isaiah 40:15) (Genesis 1:6-8)*, further confirming that the universe was created for the purpose of the humanity, and not humanity resulting from some freak accident of nature.

The Bible says that the space has a rigid framework—supported by pillars the same way we support heavy framework in our earthly structures. Here is that passage from the Bible: *"The pillars of the heavens quake, aghast at his rebuke." (Job 26:11)*. And this quaking of the pillars of the space is a reference to the cataclysmic explosion of Day 4 of creation *(Genesis 1:14-19)* which rocked space, and reverberated throughout its expanse; bringing the universe to life.

Nature was created to give humanity support and stability, not because of our value and contributions in the larger scheme of things—because we do measure up to anything—but because of God's love for mankind. The entire universe was designed to provide stability, and safety to the human race.

Just as our sun powers the earth and all the planets revolving around the sun; the glory of God directly powers our universe and causes the universe to rotate before God. Just as is the case with all the planetary systems inside our universe, <u>one full rotation of the universe</u> before God is **"one universe day"** before God. And "one universe day" is equivalent to **"365,243 earth days"** *(2 Peter 3:8)*.

Figure 1: *The Universe fits into God's mighty hand; and rotates in God's hand, suspended over nothing. (Isaiah 40:12)*

In essence, it takes the universe 365,243 earth days to complete one full rotation before God. The sun is placed in the middle of our solar system to power our world. The earth is surrounded by other planets in our solar system, some to help absorb the enormous energies that radiate from our sun, and some to deflect or swallow up space debris that come our direction from out of space.

Just like God said in the Bible that hail is reserved for the day of war, the space debris is also reserved for the day of destruction of the world and the universe. In the meantime, planets like Jupiter stands guard to prevent a lot of the space debris from reaching the earth. The rest of the planets in our solar system help to balance the magnetic field around the earth and ensure the proper workings of the planet and maintain the delicate balance it requires to maintain life in it.

Nature was design by God to reassure mankind that we are in good hands—literally. What we know remains the same to encourage us to explore the rest of our surroundings and help us find God and develop the right relationship with Him. Nature was not designed with stability to start mankind on a wild goose chase that has mired the world into sin and evil pursuits. Nature was designed by God to help mankind understand the love of God, His beauty and His majesty, and to encourage mankind to work hard to return to God and His glorious majesty.

The universe was designed by God, not only to support the solar system in helping maintain the delicate balance required on the earth to sustain life, but also to serve as signs to mankind for those things that God has not revealed to us and attract us to go in the direction of godliness and God's grace. God created the universe on Day 4 of creation by cataclysmically transforming the light He dawned on Day 1 into expansive balls of fire and energies and scattered them in various clusters all around space to create the universe, fitting them into the countless orbital highways and configurations He had created on Day 2 of creation.

Our universe, in its unimaginable size and diversity, was created by God to separate His eternal kingdom—God's dwelling place—from the earth and humanity. God knew that mankind would fail because of who we are, so He created between Himself and us a formidable barrier which He plans to lift off at the end of time so His eternal light could grace the lives of those who survive the destructions of the end-time.

God does not call Himself the "Most High" because He likes big titles like the humans do. God calls Himself the "Most High" because God, indeed, is the most high of everything and everybody that exists in heaven, in the universe and on the earth. God sits enthroned at the highest point of everything there is. His infinite size dwarfs everything He had created. And with one hand, He can clamp down on the entire universe in one giant squeeze, and make everything vaporize in less than a second.

The Bible says that Jesus Christ ascended higher than all the heaves so He could fill the entire universe *(Ephesians 4:10)*. And this is not figuratively. It is in real terms. The ancient people could not wrap their minds around a being that big that the being literally fills the entire universe. Therefore, they thought this statement was a metaphor.

And most of the people in our world today still have problem conceptualizing this. But the God of creation is truly of infinite size, infinite power, infinite might, infinite authority, infinite grace, infinite glory and infinite of all that is good. This is why the Bible says that God has *"the supremacy in everything"(Colossians 1:18)*.

In real terms, the universe—and the earth inside the universe—comfortably fit into one hand of God, on which they sit, suspended over nothing. Si if the entire universe fits into one hand of God, how big, therefore, is God? Your guess is as good as mine!

And because God has supremacy in everything, He made His Son, Jesus Christ, have supremacy in everything as well—*"God was pleased to have all his fullness dwell in him." (Colossians 1:19)*. Here is the rest of the passage from the Bible:

*"The Son is the image of the invisible God, the firstborn over all creation. ¹⁶ For in him all things were created: things in heaven and on earth, visible and invisible, whether thrones or powers or rulers or authorities; all things have been created through him and for him. ¹⁷ He is before all things, and in him all things hold together. ¹⁸ **And he is the head of the body, the church; he is the beginning and the firstborn from among the dead, so that in everything he might have the supremacy. ¹⁹ For God was pleased to have all his fullness dwell in him,** ²⁰ and through him to reconcile to himself all things, whether things on earth or things in heaven, by making peace through his blood, shed on the cross." (Colossians 1:15-18)*

9

God does not own the earth and the universe and everything in them in the same sense the human beings own things. The humans can own things that are much larger than them, for instance buildings and lands. Humans are only able to retain their rights of ownership through legal means afforded by their respective centralized governments. Humans have to turn to the legal systems to fight off anyone intruding into their properties or property rights.

One man can own a land that is as large as Manhattan and maintain legal rights over the land but would not be able to fit the entire land in his hand and clinch his fist over it. Nor could he, in real time, see what happens in every part of the land that he owns that is the size of Manhattan. But God can, and does! That is why He easily and masterfully oversees a creation as large as the universe, and at the same time monitors and controls everything inside of it, to the minutest details.

Everything God tells us in the Bible that He owns, are collectively much smaller than God in size that God has them in His hand. That is why God constantly reminds us in the Bible that nothing can measure up to Him in any way, form or shape. He is the ultimate being in any way imaginable. And this is why God says that apart from Him, there is no God.

Nothing can be compared, with any meaningful significance, to the God of creation. The collective size of everything that exists on the earth and in the universe is negligible as compared to the size of God. And this is why God has to conceal Himself from our view—with the universe—so that we would not obey Him out of fear because of His sheer size and might; but that we may obey Him out of love, because we care.

Owning things by legal rights and maintaining them through government oversight is the reason why 1% of the world's richest people own about 50% of the world's entire wealth. But ownership by the Almighty God literally means total containment, absolute control, and precise supervision and monitoring of every detail, small or large; plus active and decisive intervention when necessary. That is why the mighty and infinite God has the world and the entire universe in one hand and yet is still able to keep track of each hair on my head and know when it falls off.

Everything the Bible tells us is real and scientific. All the Bible claims are precise scientific truths. There is nothing imaginary about the word of God. A billion years from now, if God had not reeled humanity in, there would still be a great number of God's mysteries the world would still be struggling to understand due to our very limited intelligence and knowledge. All knowledge comes from God to us. We uncover something only when God gives us a method to decipher the information about that thing.

The world has grown used to claiming that someone discovered this and that. No human being discovers anything, because every single human being is operated by God in real time! And anything that comes from any human being in the form of knowledge is a gift to the world, from God, through that human being. And there is absolutely no exception to this rule.

In the ancient times, a great deal of what the Bible says appeared beyond reason because mankind had not acquired the knowledge and the tools to decipher the information. In the time of Moses, Aaron and the ancient Israelites, God did not tell them that diseases were caused by microscopic organisms which are alive and swarm all over whoever was infected with them, because the people of that time could not even bear that truth since they had no way of detecting these organisms and containing them.

So God simply taught Aaron and Moses how to differentiate between infections by these microorganisms and other minor ailments. God also taught them to use isolation and quarantine to stem the spread of these microorganisms within the population;

and how to rid these microorganisms from the infected people and reunite them with the rest of the Israelite population.

Revealing the details of these microbial infections to these men with their limited knowledge had the potential to mess up their minds, and make them run away from anyone that showed any signs of infections. In such a situation, all infections would automatically had become a death sentence of all victims since no one would want to go near them; including the priests who were charged with the responsibility of handling infections.

The recent Ebola virus infection outbreak in West African may give anyone an insight into what I am describing here. As advanced as our current world is in the detecting, managing and treating microbial infections, fear of contracting the Ebola virus is still huge within the medical community. So imagine what it would have been in ancient Israel if God had told them that microscopic living organisms were responsible for the infections that were going on within their community.

These days, the world has advanced knowledge of microbial infections; advanced knowledge of managing microbial infections (the world even breads pathogens and use them in war fares); and advanced methods of preventing and treating microbial infections. And because we do, what was once mysterious has now become a common knowledge, and has even misled some into believing that if we were able to demystify microbial infections (plagues), it means that God has nothing to do with it all along. Wrong!

Nothing that happens anywhere on the earth, the universe or even in heaven escapes the eyes of God. When we, the human beings, look into space and see a distant object, we see a shadowy image that is greatly blurred. But through rigorous training and experience, we somehow determine what it is we have seen.

God, on the other hand, does not have that kind of problem. When God looks at us from His great distance away from us, He sees even the details on us that our own trained eyes could never capture. God even sees the information stored in our thoughts while we are forming them in our minds. And He sees every bit of them with the greatest clarity. God put other animals around us that are better developed in one sense or another than us, in order to demonstrate to us that what we have received is not the most specialized in everything, and as such have some restraint.

When God decided that the world was ready to learn about the microorganisms, He led a textile merchant to peer into that elaborate but hidden world of the microbes. God was sure to lead a regular man into that discovery; and not one of the trained men who had spent a great deal of their life studying diseases and their cure. God also did not give that privilege to those others who had spent years studying biology.

Nevertheless, when this novice of science first noticed the existence of these microorganisms, he realized that he had stumbled onto something revolutionary. He looked at many different samples from different sources, compiled his observations and took his compilation to the professional scientists who hailed his work as ground-breaking and outstanding.

The world has finally been brought into the mystery of microorganisms and the diseases they cause. Yet the world has not been able to eliminate these nuisance beings because their creator did not create them to be exterminated through the will of man.

The more the world advances in its collective knowledge, the more God reveals to the church in the Bible through the world's advancement. Knowledge is required to understand the word of God. All knowledge comes from God and that is without exception. I am not just talking about the knowledge of theology. I am talking about the knowledge of everything. If it is knowledge, it comes from God. There is absolutely no other source for knowledge anywhere! Everything that science and astronomy, and all of the other fields of study, has uncovered was given to the world by God. No human being in this world has the capacity to produce anything. It is all a gift from God. *(Acts 17:28)*

There is no knowledge anywhere in the world that contradicts the information God has deposited for mankind in the Bible. If any knowledge anywhere appears to be in contradiction to the Bible, someone has deliberately, or out of ignorance, incorrectly positioned that knowledge to appear to contradict God—in which case it is no knowledge at all. The knowledge of microorganisms for instance was given to mankind not so that mankind would use that knowledge to destroy faith in God, but rather to help mankind strengthen its faith in God.

Unfortunately, out of ignorance, many in the church over the centuries, had tried to change the word of God to match whatever evidence science and astronomy had uncovered because in their ignorance of the God they serve, they thought there was something incorrect about the word of God.

If there were anything incorrect about any portion of the Bible, no matter how small or insignificant, the whole Bible will cease to be the word of God. But that would never happen, because God protects His name and His word. And He also protects those who serve Him genuinely and makes only the truth come out of their mouth in their service of Him. Listen to God proclaim that in the following passage:

"This is what the LORD says—
your Redeemer, who formed you in the womb:

I am the LORD,
the Maker of all things,

who stretches out the heavens,
  who spreads out the earth by myself,
[25] *who foils the signs of false prophets*
  and makes fools of diviners,
who overthrows the learning of the wise
  and turns it into nonsense,
[26] **who carries out the words of his servants**
  and fulfills the predictions of his messengers. *"* *(Isaiah 44:24-26)*

Human being must wait for God to lead them into an understanding of any portion of the Bible; and not change any wording in the Bible in any way or form. That is meddling with the truth of God, because the word of God is the only truth there is. The compulsion felt by some within the church's leadership over the centuries to make the Bible more correct is what led to the church of God on the earth having so many differing versions of the same truth.

And the opponents of God have tried to use this fact to discredit the Bible and God entirely. Apostle Peter did not have a different interpretation of the truth from Apostle John or any of the other apostles, including Apostle Paul. Instead with one voice they authoritatively proclaimed the truth of the gospel and transformed mankind with it.

That said, even with all the meddling and the unnecessary embellishments in the hundreds of the different versions of the Bible available in our world today, God still reveals the truth to each and every individual who comes to the Bible genuinely looking for the truth. That is the power of God and the unchangeable power of His word. People may meddle with the text, but the truth of God can never be affected. And this is a promise from God.

So, from one generation to another, God reveals more of His mysteries to the world, either to the church directly or through the efforts of dedicated and hard-working knowledgeable citizens of the world. God does this because the humanity matures with His each additional revelation. As the world matures, it becomes ready for the next stage of advancement with new knowledge from God.

The way God had the Israelites go to Egypt to develop into a self-sustainable nation full of knowledge, power and wealth, is the same way God still makes His faithful servants in the church to use the gift of knowledge He had given to individuals in the secular world to firm up their faith. The gift of God is always, first and foremost, for the benefits of His faithful servants in the church, regardless of who God had brought that gift through.

Every gift of knowledge which God had given to the world at large is for the furtherance of God's kingdom on the earth. And this is why the church must embrace knowledge of all kinds, and not just knowledge of theology. I would say it again. All knowledge comes from God for the benefit of all mankind, regardless of who God gave it to—pastor, scientist, astronomer, lawyer, a child, an uneducated man or woman, etc. God gives to whoever He chooses to give to.

The pioneers of science and astronomy were not brighter than the men and women of faith whose obedience to God, and the dedication of their lives to save all mankind, paved the way for science and astronomy. The pioneers of science and astronomy were compelled by God to venture into their respective areas of interest and pursue them relentlessly for the next stage in God's plan for the world.

The world received great and improved products and services from their work but the world also was put on a detour that got the world on its current path of destruction. This fits right into the prophecies of God in the Bible, demonstrating once again that God will always be in charge of His world and His universe.

In the 1920s, God led the world into the revelation of the Big

Bang—which denotes that cataclysmic explosion that dispersed energies throughout space to form our universe. Although God gave this revelation to a priest of the church, the world seized on it and before you know it, the world made that revelation anti-God. The world now claims that God had nothing to do with the creation of the universe; that the Big Bang was an event created by chance.

The world claims that many other universes were possible but ours just happened to be the one that popped out. And, that through billions of years of the same magic, our earth became a planet and magically produced life of various forms and sophistication on its own.

From the very time God inspired the Book of Genesis, God was sure to document in that Book the various things He did on each day of creation from Day 1 to Day 6. And God did this out of His own goodness and not because He needed to persuade any human being that He really was responsible for the creation of everything. God does not need the human praise for anything whatsoever. He asked that we praise Him for our own benefits. In *Genesis 1:14-16*—in that very short passage—God recorded the Big Bang and all the other events of Day 4 of creation, including the various purposes for the various creations of that day that constitute our present day universe.

But when the claim for the Big Bang, as the mode of formation of the universe came out, and the priest who proposed the Big Bang was up in arms about the Pope's public use of the revelation to solidify the church's claim of legitimacy; the church lacked the capacity to independently tie this claim to a particular passage in the Bible so the church let it go.

The world, which vehemently rejected this claim at first because it has the hallmarks of God written all over it, eventually accepted the claim and rode its glory for its own benefits: Space exploration exploded, especially with the advent of the Hubble Telescope which peers deep into space and revealed a lot of previously unknown mysteries. Through mathematical manipulations, many theories emerged which closely approximated real life events to help these scientists and astronomers to further alienate God and ridicule the word of God.

And through many decades of intriguing and wondrous results from science, even the church joined in in celebrating science as the best thing that has happened to mankind. With the whole world singing the praises of science and astronomy, the church felt like the only way to not lose control of the gospel that had been entrusted to it was to start meddling with the word of God to align it with the facts science has produced. Faith then becomes synonymous with having the scientific truth in conjunction with retaining the label of Christianity. It becomes a commonplace to hear people talking about two ways of getting to the truth—faith and science.

Unfortunately for the world, science is not a way to salvation; only faith in God is. The facts of science can never lead anyone to salvation. So, by God demanding that we believe even when we have not seen any worldly or scientific evidence, does it mean that God is trying to keep us in error? That is preposterous! Who is the truth, then? Isn't God the source of all truths? Why would truth itself want anyone to live in lies?

That we have advanced in knowledge through science do not make us any superior to the ancient people who did not have our kind of knowledge and yet trusted God entirely to live their lives in prosperity. No scientist today could match the knowledge and understanding of Abraham or any of the other faithful people in the Bible.

Yet Abraham and the rest of these faithful men and women trusted God because they knew His word is trustworthy. And God came through for them. Science has actually damaged the faith of so many men and women in the world who, ultimately, would pay the price for looking for assurances that are no assurance at all. Only faith in God produces real assurance.

Faith in God does not preclude relying on the facts of science to live our lives. As a matter of fact, anyone who has a strong faith in God quickly realizes that all facts come from God and can be relied upon. Faith in God would never convince any believer that 1+1=3 because God does not want anyone believing that. 1+1=2 is a mathematical fact—a fact established by God to give mankind stability in its outlook in life.

But mathematical or scientific facts do not lead to faith. Facts are already well established knowledge and so preclude the need for faith. The Bible says, *"Now faith is confidence in what we hope for and assurance about what we do not see." (Hebrews 11:1)*

For a while it looks like the church is losing the battle of legitimacy against science, but God has decided to change it all. While the world has single mindedly pursued science, the church had decided to leave science to the world, until now.

Science is for all mankind, and is designed by God to provide His church, and the believers, with the confirmation of God's truth, for the strengthening of their faith. God designed science to serve a more vital function for the church than it serves for the world. But the church has so far missed that fact and had more or less turned its back on science and astronomy.

A church that stops acquiring knowledge cannot grow as God intended, because knowledge dismisses ignorance and firms up the truth.

It may be hard to believe, but the prosperity of the ancient Egypt was more for the people of ancient Israel than it was for the ancient Egyptians. God suddenly brought that civilization to its knees in order to put Joseph—a slave and a prisoner—in the highest position of authority in the civilization so that God could bring prosperity to the house of Abraham as God had promised Abraham prior.

God brought Abraham's clan into the kind of sophisticated and advanced economic and political system their little size could not have permitted them to put together. So God injected them into a fast-track environment for the prosperity God desired for them.

That is the same way God is treating His church in our current age. All of the world's sophistication and advancement is meant by God to benefit His church. The church lacks the sophistication to build the kind of reward for the church which God intends for the church. So, God allowed the world to overrun the church in the middle ages and advance to great heights in knowledge base and sophistication.

The church would then be thrust into that which is already well established. Therefore, the current state of the church is only a transitory one. The church is destined for glory because God always reserves for those who trust Him and obey Him, much better glory than He gives to those who disobey Him and ridicule His mighty name.

God has shown me that understanding His lessons are not about how much you already know: understanding God's lessons is about how uncontaminated your mind is.

If you have allowed your mind to be filled with clutters, you will have great difficulty understanding what God is teaching you. Your mind has to be a slave to your spirit for you to hear God and retain the lessons He is giving to you. And this is not something you can force. It is something that comes through unshakable faith in God and His every word.

The word of God is more precious than the finest gem and all the wealth in the world. The word of God is life and wisdom. And when knowledge is added to it, understanding and joy precipitate out of it. God never disappoints anyone who seeks His understanding. That is why the Bible says: seek and you will find; ask and it will be given to you; knock and it will be opened to you. It is always a charm! Try it and you will believe. That is all I did.

And finally, a little breakdown of the revelation in this book: One is that the entire universe sits in the middle of God's mighty hand. Let us look at the passage together from the Book of Isaiah—the passage that yielded this revelation. I have referenced the passage a few times before in my other books. And this time, I came to the same passage, as yet another reference to support of my discussions in this Book about science omitting the all-important "speed of God"—which is infinitely faster than the speed of light—in its estimation of the age of the universe and of the earth, and other scientific considerations.

As I was lifting this passage off the Bible, I felt it in my spirit that the passage I was copying is a scientific revelation by itself—a revelation of huge scientific implications. On a second look, the whole thing became as clear as a cloudless day in my mind. I have just received a revelation!

A day or so before this, I had been made to understand that the passage in the second epistle of Peter in reference to God and time is not simply a metaphor, but also a scientific fact tucked away in the pages of the Bible. And both of these revelations came at the heel of the "speed of God" revelation which also stole into my spirit like a thief. So, in a matter of two days, I was given three major pieces of information that are inseparably linked to the creation of the world and the universe by God. Here are all three revelations with more details:

The universe was created by God at the "Speed of God" within 48 hours—on two separate days:

The *"speed of God"* is infinitely faster than *186,000 miles per second* which is the ***"speed of light"****! Coupled with the scientific claims, these revelations from the Bible put the universe rotating at a speed that is much faster than the speed of light because the universe's rotation is directly powered by God's radiance; and not by some created star, as is the case within the universe. We will get into more details about this in a short time.*

God created the entire universe in only 24 hours—Day 4 of creation *(Genesis 1:14-19)*. Two days prior to Day 4 (that is, on Day 2) of creation, God had created space, fully equipped with all the orbits, their respective speeds, their direction of movement and their configurations and planar orientations *(Genesis 1:6-8)*.

Here is the passage from the Bible about the creation of space by God within 24 hours: the orbits; their respective speeds; their direction of movement; and their configurations; and planar orientations:

"And God said, Let there be a firmament in the midst of the waters, and let it divide the waters from the waters.

[7] And God made the firmament, and divided the waters which were under the firmament from the waters which were above the firmament: and it was so.

[8] And God called the firmament Heaven. And the evening and the morning were the second day." (Genesis 1:6-8 KJV)

And here is the passage from the Bible about God creating the entire universe within 24 hours that comprised Day 4 of creation:

"And God said, Let there be lights in the firmament of the heaven to divide the day from the night; and let them be for signs, and for seasons, and for days, and years:

[15] And let them be for lights in the firmament of the heaven to give light upon the earth: and it was so. [16] And God made two great lights; the greater light to rule the day, and the lesser light to rule the night: he made the stars also.[17] And God set them in the firmament of the heaven to give light upon the earth, [18] And to rule over the day and over the night, and to divide the light from the darkness: and God saw that it was good.[19] And the evening and the morning were the fourth day." (Genesis 1:14-19 KJV)

One full rotation of the universe before God is equal to 365,243 earth days:

God was at hand to observe the first full rotation of the planet earth, powered by the original light which God dawned on Day 1 of creation. The very instant of the dawn of the light kicked off time as we know it and started the earth on rotation. *{Revolution did not occur until God created light in the expanse of the sky and positioned our sun to take over from the original light of Day 1 through Day 3 of creation.*

The fact that the earth had only rotational motion for the first three days of its creation; and did no revolution around that original light of Day 1 through Day 3 of creation, indicates that God never allows anything to get behind Him. And throughout the Bible God, talks about hiding His face from something out of displeasure, and the troubles that result, for that thing or person, when He hides His face.

These taken together, explains why the revelation of 2 Peter 3:8 centered only on "universe day"—and not "universe year"—because it is rotation that produces the day, whereas revolution produces the year.

23

This reference in the Bible to a day that is as long as a thousand "earth years" is not a mere figure of speech. The same thousand-year-to-one-day time reference is also recorded in the Book of Psalms.

And the passage in 2 Peter 3:8 specifically said **"with the Lord",** *poignantly pointing to a time system that belongs solely to God; aside from, and in comparison to, that which man is familiar with.*

With God, everything is real because God is all about purpose, and as such, God does not do or say anything to mislead His valued children—mankind—in any way. Through subtlety, God leads humanity to huge revelations of immeasurable benefit—spiritually and in science and worldly knowledge.

Before the earth overcame inertia and started its rotational motion to kick off our current time system—and the sun and the universe came in on the fourth day of creation—time was simply a continuum without end. So will time be once again, when God brings judgment upon the world, and reabsorbs the energy currently sustaining our universe in an implosion, as prophesied in Revelation in other passages in the Bible.} Here is the passage from Second Peter:

"But do not forget this one thing, dear friends: With the Lord a day is like a thousand years, and a thousand years are like a day. ⁹ The Lord is not slow in keeping his promise, as some understand slowness. Instead he is patient with you, not wanting anyone to perish, but everyone to come to repentance." *(2 Peter 3:8-9)*

When Peter makes the comparison of time in His epistle, he was not just sounding off. He was stating a fact from God which has deep scientific implications. But because the church does not typically look for science in the Bible, and because the fact which permits even science to make sense out of this proclamation, has only recently become available. And since after the middle-ages, the last place the scientist would go to for the truth is the Bible because most of them detest the Bible.

God does not inspire empty talks. Every word that comes from

God establishes God's truth, and all the Bible was inspired by God *(2 Peter 1:20-21)*. It was with words that God created the earth and the universe and everything in them *(2 Peter 3:5-7)*. And the Bible tells us that nothing that comes out of the mouth of God returns without achieving what the word was intended by God to achieve.

God marked off the universe with the "breadth" of his hand:

In *Isaiah Chapter 40*, God establishes for mankind the size of the universe; gives detailed comparisons of the various things He created, and provided comparative sizes of the earth, its continents, the nations of the earth, and their collective size with respect to the universe. God finally concluded in this comparative analysis that all the nations of the world amounted to nothing. He went further to call them collectively, "less than nothing."

In *Isaiah 40:12*, God talked about the waters of creation, in which the earth sat in *Genesis 1:2*, being in the *"hollow of His hand"*. God then unequivocally declared that no human being has the capacity to measure the extent of the waters because of its astronomical size. Yet the water—in spite of its astronomical size—was contained within a shallow depression on the palm of God's mighty hand, with our huge earth sitting right inside the water in *Genesis 1:2*.

If the all-knowing God knew that someday in the future, He would grant mankind knowledge of science to help humanity understand God's mysteries better, why then would God claim in the Book of Isaiah that no one could possibly measure the water in the "Hollow of his hand"? Because God was, and is still, confident that man and all his advances in science and technology to the end of time could not possibly measure that water because that water in the "hollow of God's hand" is truly immeasurable.

God knew that declarations like this, in the face of the little window of knowledge He had cracked open for mankind, would inflate man's ego and get man fantasizing about bringing God down, just like Satan did, and fell. Science would look at the claims in *Genesis 1:6-8* about God, in creating space, pushing the water above the void out of space and quickly mistaken that to mean the clouds in the sky—in the immediate vicinity of the earth.

And because mankind had mistaken that water that was pushed out of space to mean the moisture in the clouds in the earth's sky, mankind— Christians and scientists alike—have dismissed that account of creation as either a metaphor or a flat-out lie, respectively. And as such, mankind had moved on with life, putting that account behind it as just another bogus declaration from the ancient "relic" that is the Bible. But mankind has greatly erred.

That account of "water being pushed out of space" in Genesis Chapter One, is scientifically accurate, and within God's infinity capacity. As a matter of fact, the Bible says that such feat is *"but the outer fringe of God's works." (Job 26:7-14)*

Psalm 95:4 also declared that the water of creation in Genesis Chapter One is physically inside the hand of God. Here is the passage from the Bible: *"For the LORD is the great God, the great King above all gods.*
⁴ In his hand are the depths of the earth,
   and the mountain peaks belong to him.
⁵ The sea is his, for he made it,
   and his hands formed the dry land." (Psalm 95:3-5)

If anyone is confused as to whether verse 4 is talking about the water physically sitting in God's hand or metaphorically saying that God has charge over the water, pay attention to the word "hand" in that verse. Hand is singular. When something being in someone's hand is used metaphorically, the plural "hands" is used. Here is one such example from the Bible: *"Jesus called out with a loud voice, "Father, into your hands I commit my spirit." When he had said this, he breathed his*

last." *(Luke 23:46)*

In this passage, Jesus Christ did not physically place His Spirit into God's hands; to the extent that any human being can determine or verify. Jesus Christ entrusted His Spirit to the care of God. When putting something into the care of another person, or giving someone the authority or privilege to decide on the fate of something, hand is always used as plural—hands.

But in all the instances where the earth, the universe and the waters of creation are described in the Bible as being in God's hand, "hand" is used as singular, because the declarations are talking about God's hand as the physical location of the earth, the universe and the waters of creation.

Isaiah 40:12 and *Psalms 95:4* are, therefore, stating scientific facts about the physical location of the water of creation in which the earth was submerged in *Genesis 1:2.*

Two-thirds of the earth's surface had been determined to be covered by water. And humanity had very much mapped out the extent of water on the surface of our earth. But the water on the surface of our earth and the moisture in the sky above the earth which gives rain to the earth, together do not account for "the water in the hollow of God's hand" that is referenced many times in the Bible.

The water we have in our seas and in the moisture in the sky above the earth is just a small fraction of "the water in the hollow of God's hand". Because, according to *Genesis 1:6-8*, God, in creating space, pushed more of the water contained in the hollow of His hand beyond the outer limits of space, thus creating space under vacuum.

Knowledge of Geometry would readily convince anyone that to draw a vacuum within a spherical void, the water on top of the astronomically large sphere called space has to be much larger than the water that remains around the much smaller inner sphere called the earth.

Since the earth inside the water in *Genesis 1:2* is infinitesimal in comparison to the space God created within the water in *Genesis 1:6-8*—that is, the water in the hollow of God's hand—we must therefore conclude that the water on top of space is infinitely larger than the water which God left on the earth to become earth's seas.

And although space was fully formed on Day 2 of creation, and the volcanic explosion that created land out of the sea completed on Day 3, God out of His own will could still choose to use water elsewhere in the universe to douse volcanic materials into solid planets, hence the proliferation of planetary systems all over the universe.

God created space on Day 2. Then on Day 3, He formed the universe through a cataclysmic explosion we now dubbed the Big Bang. In only 24 hours that constituted Day 4 of creation, God set the universe in place—*Genesis 1:14-19.*

And for the benefit of those who spend endless hours looking around the universe for signs of water outside the earth, all they need to do to find water is to look at the outer perimeter of the universe and they will blow their minds as to the deluge of water that exists there. Additionally, finding water anywhere else in the universe does not in any way refute anything the Bible says about God's creation activities. Everything man's reason produces is through guess work, conjectures and suppositions. None of which our great God has any time for. When He opens His mouth, only the finest word comes out of it; and blesses those who give credence to it.

At His own choosing, God could have let water get anywhere He so chooses in the universe. As a matter of fact, God could still be using water in the new things He talked about in *Isaiah 48:6-9.* He is God and has absolute freedom to do as He sees fit. Here is God saying it: *"I make known the end from the beginning, from ancient times, what is still to come. I say, 'My purpose will stand, and I will do all that I please.'"* (Isaiah 46:10)

And here is the Scripture that gave up its scientific secrets—now that the world's scientific advancements have enhanced humanity's ability to understand what God is proclaiming in these verses. That the entire universe and all of its contents, including the earth and mankind, is sitting in God's hand; and squarely supported, monitored and controlled by that mighty hand of God:

"See, the Sovereign LORD comes with power,
and he rules with a mighty arm.
See, his reward is with him,
and his recompense accompanies him.
11 He tends his flock like a shepherd:
He gathers the lambs in his arms
and carries them close to his heart;
he gently leads those that have young.

12 Who has measured the waters in the hollow of his hand,
or **with the breadth of his hand marked off the heavens**?
Who has held the dust of the earth in a basket,
or weighed the mountains on the scales
and the hills in a balance?" (Isaiah 40:10-12)

"Surely the nations are like a drop in a bucket;
they are regarded as dust on the scales;
he weighs the islands as though they were fine dust.
16 Lebanon is not sufficient for altar fires,
nor its animals enough for burnt offerings.
17 Before him all the nations are as nothing;
they are regarded by him as worthless
and less than nothing." (Isaiah 40:15-17)

"Do you not know?
Have you not heard?
Has it not been told you from the beginning?
Have you not understood since the earth was founded?
22 He sits enthroned above the circle of the earth,
and its people are like grasshoppers.

He stretches out the heavens like a canopy,
  and spreads them out like a tent to live in." (Isaiah 40:21-22)

""To whom will you compare me?
  Or who is my equal?" says the Holy One.
26 Lift up your eyes and look to the heavens:
  Who created all these?
He who brings out the starry host one by one
  and calls forth each of them by name.
Because of his great power and mighty strength,
  not one of them is missing." (Isaiah 40:25-26)

The breadth of God's hand is a measurement and not a measuring tool. God's two hands are the most effective and most precise tools anyone could employ in any project of any size! So when *Isaiah 40:12* states that God *"with the breadth of his hand marked off the heavens"*—the heavens being the universe—the passage is giving you a total measurement and not simply talking about a measuring tool that was used an unlimited number of times to establish a measurement.

He is God! So, He could easily give Himself whatever He wanted. It was only in primitive times that even the human beings used the breadth of their hands, the length of their foot and their strides, to make measurements.

Mankind had since moved on to tape measures, lengthy rolls of tapes, rollers with counters, lasers and other more intricate computing means like the speed of light to determine huge distances. Therefore, if human beings have been able to make this much stride in their measuring techniques, how much more could God have done—the God who created the most wonderful things that exist anywhere?

It would be timid for the awesome God who made the entire universe and the most formidable forces anywhere to resort to counting off the breadth of his hand to measure out anything He wants to build. Mankind in their improvement of their measuring techniques had to grow and mature as they went from one improvement to another, but the all-knowing God was perfection right from the very beginning and as such could not have resorted to making crude measurements. So what this passage from the Book of Isaiah was talking about is the entire measurement of the universe, and not an instrument of measurement that was counted off an infinite number of times.

Take a look at the Bible's description of God's measurement when God created the earth. Keep in mind that God created the foundations of the earth prior to Day 1 of creation; and land and sea on Day 3 of creation. So, is it possible that the same God who talked about using *"measuring line"* across the earth to establish the dimensions of the earth *(Job 38:5)*—which is very tiny in comparison with the expansive universe—would suddenly resort to a much cruder method of measurement in establishing the dimensions of the universe which is infinitely larger and more complex than the earth? Impossible! Our God does not regress. He is perfection, through and through! Here is the passage from the Bible:

"Where were you when I laid the earth's foundation?
  Tell me, if you understand.
⁵ Who marked off its dimensions? Surely you know!
  <u>Who stretched a measuring line across it?</u>
⁶ On what were its footings set,
  or who laid its cornerstone—
⁷ while the morning stars sang together
  and all the angels shouted for joy?" (Job 38:4-7)

Therefore, *Isaiah 40:12* establishes that the entire universe was designed to fit into God's mighty hand and be eternally supported by His hand. And that being the case, how big then is our God? This finally paints a picture for mankind about the unimaginable size of our God— the God who has chosen to make us His children and shower us with His endless love. Not only did He build a special place called the earth for us, enamored with everything necessary to keep all mankind in safety, He also went out of His way to put us in His eternal will, not because of our own self-importance because we measure up to nothing; but because of His great love for mankind.

So when *Isaiah 40:17* proclaims as follows: ***"before him all the nations are as nothing; they are regarded by him as worthless and less than nothing,"*** it was not God talking about mankind in a derogatory way because He was angry at mankind.

Calling all the nations of the world, collectively, nothing was simply God establishing a scientific fact; because in the arithmetic of scales, all of humanity; and all the nations of the world and all they possess; physically amount to nothing.

And truer still, mankind, and all, amount to less than nothing; and God was sure to state it that way. This further establishes the fact that God is doing everything that He is doing with mankind out of His immense love for mankind, and not because mankind amounts to anything. Why, then, do we have difficulty reciprocating that love by obeying the commands of God?

Additionally, we can see why nobody has been able to see God. God placed the humongous universe between Himself and our beloved earth—the same earth which we are unfortunately bent on destroying out of greed.

Also, with God seated at His throne outside our universe and being as big as He is—with the entire universe sitting on His hand—no human effort could get us to where God is. And nothing we do on this earth could affect Him in any way, shape or form.

Only our obedience to His commands would achieve anything for any of us. And it is not too late to start now. The new revelation contained in this book is extremely bad news for those who are in great hurry to remove the name of God in everything in our society. At the same time, it is extremely good news for those who desire God's love and obey His commands. I am personally elated and all mankind should be, too.

Look at the following passage from the Book of Job *(Job 26:7-14)* for more details on God's creation activities. Also see the inserted references from Genesis Chapter One:

Job 26:7-14

"He spreads out the northern skies over empty space; -- {Genesis 1:6-8}

he suspends the earth over nothing. – {Genesis 1:2}
⁸ He wraps up the waters in his clouds,

yet the clouds do not burst under their weight. – {Genesis 1:2}
⁹ He covers the face of the full moon,

spreading his clouds over it. – {Genesis 1:14-19}
¹⁰ He marks out the horizon on the face of the waters

for a boundary between light and darkness. – {Genesis 1:2-5}
¹¹ The pillars of the heavens quake,

aghast at his rebuke. – {The "Big Bang" (Genesis 1:14-19) resonated across the space God created on Day 2 , Genesis 1:6-8}
¹² By his power he churned up the sea; – {The volcanic eruption that created land out of the sea and the ensuing tsunamis that rattled the sea and weathered the land Genesis 1:9-10, for the creation of vegetation on the land Genesis 1:11-12}

by his wisdom he cut Rahab to pieces.
¹³ By his breath the skies became fair;

his hand pierced the gliding serpent.
*¹⁴ **And these are but the outer fringe of his works**;*

how faint the whisper we hear of him! (Job 26:7-14)

And to support the creation account of Genesis Chapter One about the earth being created before the universe, look at the following passage from the Book of Proverbs. *Proverbs 8:23* declares about wisdom—the first of all of God's works—*"I was formed long ages ago, at the very beginning, when the world came to be." (Proverbs 8:23)*

Let us break it down mathematically so we can gain a better understanding of what the passage *(Proverbs 8:23)* is saying:

"Long ages ago"= before all creation (as can be seen in Proverbs 8:24)

And,

At the very beginning = When the world came to be

It does not get any clearer than that. The beginning of the world (earth) was the beginning of the earth, the universe and everything in nature. And the writer of Proverbs was sure to take us through the entire creation in the same very order things happened in Genesis Chapter One.

Following this opening statement in verse 23 about the very beginning—when the world came to be—verse 24, 25 & 26 recount the watery depth of "Before Time" *(Genesis 1:2);* then on to the springs, the mountains, the hills and the fields of Day 3 of creation *(Genesis 1:9 & Genesis 2:6)*.

Verse 27 & 28 then talk about the creation of the universe *(Genesis 1:14-19)*, the marking off of the horizon on the face of the deep for boundary between light and darkness, the establishment of the cloud in reference to the moon and the other celestial bodies, and the establishment of the clouds above the earth for rain cycles for the replenishment of the sea (referred to in the passage as the securing the fountains of the sea).

God holding the entire dust of the earth in a basket indicates how really small the earth is as compared to some significantly larger creation of God that is referred to here as a basket. God weighing the mountains of the earth on the scales, and the hills of the earth in a balance, indicates that God knows exactly how much they collectively constitute as measured against everything else God created.

But the ancients had no way of knowing that these declarations are literal and not metaphors. The ancients could see the sky but had no way of knowing about the existence of other planets in our solar system, much less the nature of our expansive universe. But now, science and technology have provided mankind with the knowledge that permits us to put these things in perspective as we are doing in this book.

And this should only be the beginning. Theology has to embrace the knowledge of science and astronomy for a better understanding of God's truth in real terms.

God then talked about all the nations of the earth, collectively, as amounting only to a single drop into a bucket. And what is that bucket in which this drop is dropped into? The universe!

Therefore, the earth and all its inhabitants and their collective wealth and civilization, as compared to a huge bucket that is the universe, amounts only to the size of a tiny drop of water set in the middle of the expansive universe.

The insignificant size of the earth and all of its nations and their peoples and wealth and civilization, are regarded as a single particle of dust on God's scale of measurement, which God uses to measure all created things, up to and including the universe itself. And these measuring capabilities are all within God. God does not need tools to create. This is why God forbade His people never to use any tools in constructing anything dedicated to God. *(Exodus 20:25 and other places in the Bible)*

All the continents on the earth are described in *Isaiah 40:15* as "islands" because they appear that way from space, let alone from God's vantage point—higher up beyond our universe. And these continents put together are regarded by God *"as though they were fine dust"*. That is, taken together, they appear as though they are a minute particle of dust.

Isaiah 40:16 brought that comparison even closer for those who in ancient times did not have our knowledge of space and the universe. The passage talked about the entire country of Lebanon with its extensive priced timber not being enough for one alter fire; and all the animals in the ancient lush wilderness of Lebanon not being enough for burnt offerings to God at His throne in heaven.

These descriptions are also meant to convey to humanity that nothing we can offer to God, out of our individual wealth, could meet God's criteria of a sacrifice in any way. That what God demanded from the ancient Israelites for sacrifice, and still demands from believers today, is a token, designed by God to compel everyone to show his or her gratitude to God on a regular basis.

This passage from Isaiah concluded by declaring that all of the earth, its people and their collective wealth and civilization, measure on God's scale of measurement as nothing—or more accurately put, less than nothing. Is that assessment not truly a good comparison between the earth and everything in it, including mankind, to the rest of the expansive universe? On a real scale of measurement, mankind and everything mankind brings to the table equates to less than nothing and could easily be discounted without anyone noticing that something is missing.

And according to Physics, *"The size of dust particles varies from about half a micrometer (0.00002 in) to several times this size."* *(http://hypertextbook.com/facts/2003/MarinaBolotovsky.shtml)*

In more familiar terms, 0.00002 in or 0.5 micrometer is 2/10,000 (two portions out of ten thousand divisions of one inch). This is the size of our entire earth from this comparative analysis by God, of our earth with respect its importance, on a scale on which the universe is also weighed.

Below is an internet excerpt, describing the different stages in the formation of a rain drop. God makes the rain and when the passage in Isaiah talks about all the nations of the world collectively being like a drop in a bucket, rain drop could not be far from the drop the passage was referring to. So, let us look at the various sizes of rain drop and their stages of formation from this short excerpt:

"Why raindrops are different sizes

In science we learn that one question often leads to another, or several others. Before we can discuss raindrop sizes, we must understand what a raindrop is. How is a raindrop made? How big can a raindrop be?

In order to have rain you must have a cloud--a cloud is made up of water in the air (water vapor.) Along with this water are tiny particles called condensation nuclei--for instance, the little pieces of salt leftover after sea water evaporates, or a particle of dust or smoke. Condensation occurs when the water vapor wraps itself around the tiny particles. Each particle (surrounded by water) becomes a tiny droplet between 0.0001 and 0.005 centimeter in diameter. (The particles range in size, therefore, the droplets range in size.) However, these droplets are too light to fall out of the sky. How will they get big enough to fall?

Picture a huge room full of tiny droplets milling around. If one droplet bumps into another droplet, the bigger droplet will "eat" the smaller droplet. This new bigger droplet will bump into other smaller droplets and become even bigger--this is called coalescence. Soon the droplet is so heavy that the cloud (or the room) can no longer hold it up and it starts falling. As it falls it eats up even more droplets. <u>We can call the growing droplet a raindrop as soon as it reaches the size of 0.5mm in diameter or bigger. If it gets any larger than 4 millimeters, however, it will usually split into two separate drops.</u>

The raindrop will continue falling until it reaches the ground. As it falls, sometimes a gust of wind (updraft) will force the drop back up into the cloud where it continues eating other droplets and getting bigger. When the drops finally reach the ground, the biggest drops will be the ones that bumped into and coalesced with the most droplets. The smaller drops are the ones that didn't run into as many droplets. Raindrops are different sizes for two primary reasons.

1. *initial differences in particle (condensation nuclei) size*
2. *different rates of coalescence.*

This information is courtesy of the University of Idaho."
(http://water.usgs.gov/edu/raindropsizes.html)

In conclusion, rain drop ranges from 0.5 millimeter to 4.0 millimeter. Which particular size the passage in Isaiah was alluding to we do not know but is certainly between these two sizes.

Half a millimeter is equal to 0.0197 inch (that is approximately 0.02 inch, which is 1 portion out of 50 divisions of one inch.)

And bucket sizes are wide and varied. So with the bucket in this narrative representing the universe, and the drop representing the earth, with all of its inhabitants and their earthly possessions and knowhow, the passage's comparison hits the point home.

This comparison was designed to convey that the universe, and its contents and treasures, far outsizes the earth, and all of its contents and treasures.

Therefore, *Isaiah Chapter 40* was designed by God to give humanity a good reveal into the science of everything. Up until now, this chapter of Isaiah and others similar to it in the Bible had sounded like an angry rebuke from God to mankind, when in reality it is all scientific facts with the highest precision—designed by God to give mankind unobstructed peek into God's creations, His love for mankind and His wonderment; so that mankind would be persuaded into readily becoming compliant with the commands of God in the Bible.

What God had designed to show love, has been mistaken for angry talk by the world, and has been used by the majority of the world to bash

God and defame His mighty name. And while this was happening, humanity then ran off into uncharted territory, cooking up lies and masking its lies with exotic mathematical formulas and unfounded theories; and has been receiving glory for them.

Following is a the passage from Proverbs Chapter Eight, part of which was referenced above to further support the order of creation as God laid out in Genesis Chapter One:

Proverbs 8:22-31

"The LORD brought me forth as the first of his works,
before his deeds of old;
23 I was formed long ages ago,
at the very beginning, when the world came to be. {YES! The world
began it all}
24 When there were no watery depths, I was given birth,{ Genesis 1:2}
when there were no springs overflowing with water;{Genesis 2:6 &
Genesis 1:9-10}
25 before the mountains were settled in place,{Genesis 1:9-10, Job
38:14}
before the hills, I was given birth,{Genesis 1:9-10 & Job 38:14}
26 before he made the world or its fields{Genesis 1:9-10}
or any of the dust of the earth.{Genesis 1:2}
27 I was there when he set the heavens in place,{Genesis 1:14-19}
when he marked out the horizon on the face of the deep,{Genesis
1:14-19 & Job 26:10}{the light source in Genesis 1:3-5 was God
Himself. His radiance, while seated on His throne in His dwelling place,
directly illuminated the surface of the deep as the clouds resulting from
the quenching of the volcanic mass (earth) God dropped inside the water
in the hollow of His hand dissipated .The light covered half of the
earth/water at a time to establish the horizon over the waters—that line
between light and darkness}
28 when he established the clouds above{Genesis 1:14-19 & Job 38:9}
and fixed securely the fountains of the deep, {Genesis 1:9 & Job
38:8,10,11}
29 when he gave the sea its boundary{Genesis 1:9 & Job 38:8,10,11}
so the waters would not overstep his command, {Genesis 1:9 & Job

38:8,10,11}

and when he marked out the foundations of the earth. *{Genesis 1:2}*

[30]   **Then I was constantly at his side.** *{Wisdom belongs with the TRINITY}*

I was filled with delight day after day, *{Wisdom belongs with the TRINITY}*

   rejoicing always in his presence, *{Wisdom belongs with the TRINITY}*

[31] **rejoicing in his whole world** *{Wisdom belongs with the TRINITY}*

   and delighting in mankind. *{Wisdom belongs with the TRINITY}*

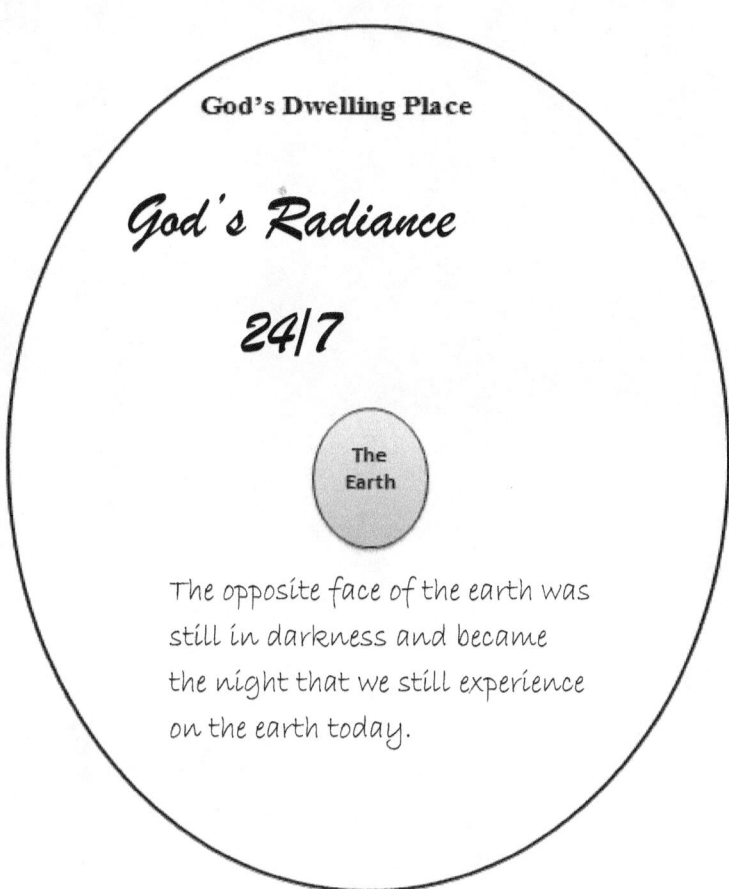

Figure 2: Heaven & Earth Configuration at Genesis 1:2-5 *(Not Drawn to Scale). God's glorious radiance filled the entire God's dwelling place (heaven) and; to a modified capacity, the surface of the earth/water, to create the day and night configuration that God desired to create on the earth, and maintained on the earth through the end of time; at which point the earth will be completely covered 24/7 (without nights) and into eternity, with God's direct radiance.*

Ifeanyi Chukwujama

Chapter 2

Our Universe is a physical between Earth and Heaven!

Our Universe is the Great Divide! It is an indomitable physical barrier intentionally placed by God between our earth and heaven—God's dwelling place. When God first created the earth, there was no space and there was no universe *(Genesis 1:2)*. There was only our mighty God building a safe and secure place for mankind out of the waters that were sitting in the hollow of His hand *(Isaiah 40:12) (Genesis 1:2-5)*, while the heavenly hosts marveled and rejoiced at His majestic creation *(Job 38:7)*. See Figure 2 below, and compare it to Figures 3 & 4.

Everything recorded in Genesis Chapter One, and the order in which they are recorded—from Day 1 to Day 6—is the truth and the only truth about creation of everything on the earth and in the universe. Every other version of creation you have heard from every other source is a fallacy, and a figure of someone's imagination.

The widely accepted scientific version which says that the universe came into existence before the earth is a flat-out lie—a lie that is dressed in fabricated theories, fancy mathematics and science fiction. It is not everything that comes from our so-called science that is science.

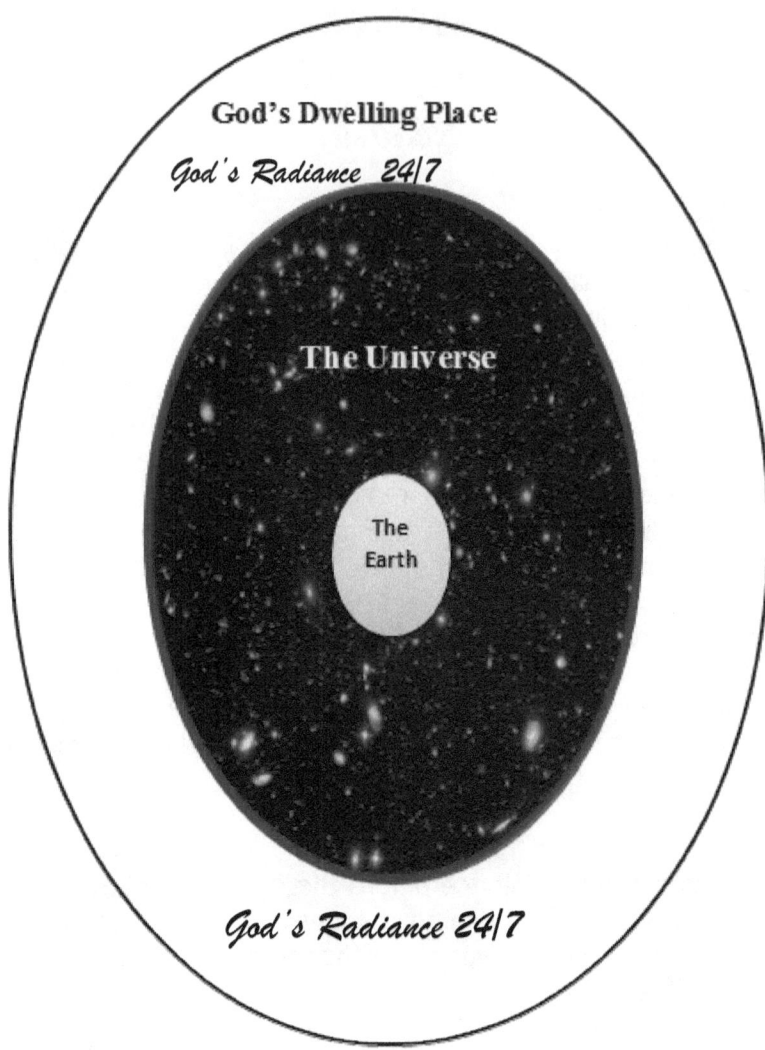

Figure 3: Present-Day Heaven, the Earth and the Universe Configuration *(Not Drawn to Scale). The universe is wedged between the earth and the universe to completely block earth's view of heaven—God's dwelling place, thus denying the earth of God's direct radiance.*

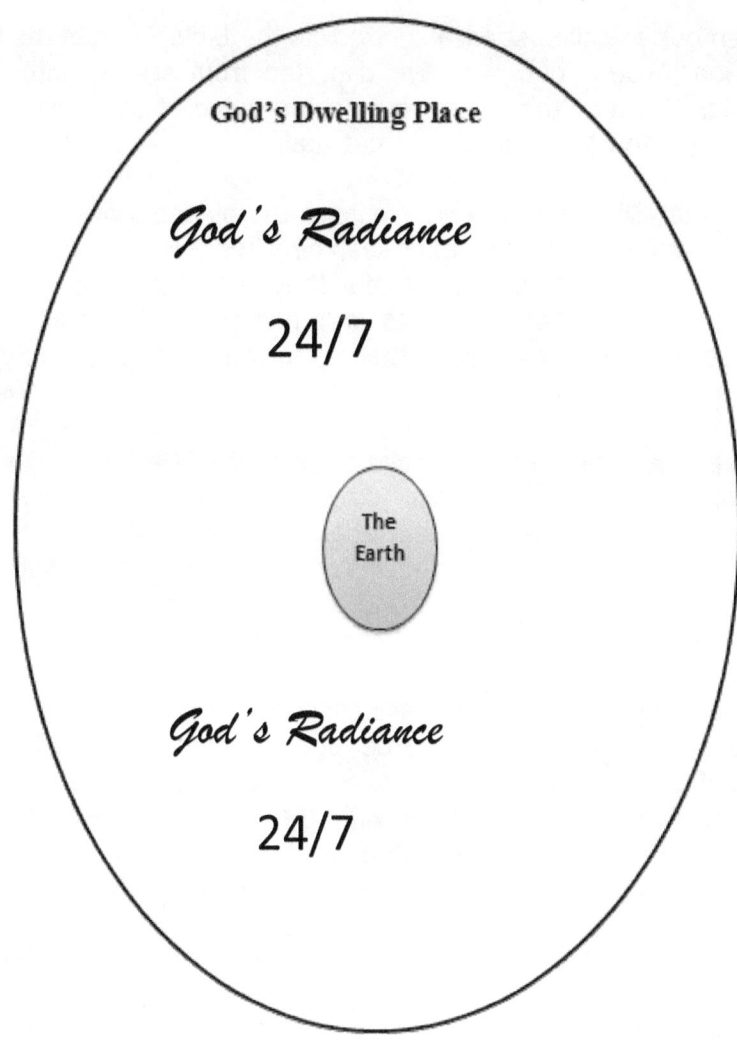

Figure 4: End-Time Heaven & earth Configuration *(Not Drawn to Scale). The universe has imploded and all of the energies that currently exist in the universe have been reabsorbed by God, as prophesied in the Book of Revelation and other passages in the Bible.*

Science is a creation of God! Therefore, God knows science best because He formulated every aspect of it, and applied it in all His works. And science completely affirms the truth of the Bible when we get it right *(Proverbs 8:1-36)*.

When our so-called science contradicts the Bible, or puts the truth of God in bad light, we have departed from science into black magic. The only thing that could come out of that is trouble for mankind. And the Bible guarantees that.

The great Bible did not only create in the pioneers of science the inquisitiveness that got them searching for answers, it provided them with road maps to the things they were curious about. That is the reason all early scientists were part of the Christian faith or the Jewish faith or religions that borrowed from Christianity and Judaism.

And as they followed these road maps that still exist today in the Bible, they made the great discoveries they are known for. Unfortunately, however, some abandoned the lead of the Spirit that got them the glory and instead pushed farther than they were allowed by the Spirit of God. And the results were erroneous conclusions that began to taint true science.

And as they came up with these conclusions that negate the truth of God, the church pushed harder against them, further alienating itself from this emerging fledgling group of distinguished men. Because of their displeasure with the church's censorship and treatment, their collective sentiment against the church paved the way for emerging political powers in Europe, and later America, to snatch them up and form the union between science and the government that we see everywhere in the world today. And the church has been losing power and influence since then—now to the point of ridicule by the masses in many developed nations that benefited the most from this union between science and the government.

And out of pressure from the popularity of science, and its success in the new union with government, some Bible teachers, preachers and scholars have mistaken the thesis statement of *Genesis 1:1* as meaning that God had created the heavens (that is the universe) before He created the earth.

That assumption or belief is not only incorrect but a distortion of the word of God as well. The people of God must remain calm at every point in their lives because the God they serve is mighty beyond everything. They must always wait for God to lead the way; and not make up things to counter an attack by the enemy.

The experiences of the ancient Israelites must always be a guide for the Christians in our world today. God sometimes allows the enemy to gain in on God's people *(Revelation 13:7-8)*, so God can test the faith of His people. God also does it to embolden the enemy some more, so that when God finally intervenes, the enemy's defeat would become a great and tumultuous victory for the people of God.

We are in the end-time, and had been since Jesus Christ rose from the death and ascended to the Father in heaven. Here is the Bible telling us that the enemy of God was given power By God to smear God's people and destroy them, so that the enemy can pressure them into turning away from God:

"The beast was given a mouth to utter proud words and blasphemies and to exercise its authority for forty-two months. ⁶ It opened its mouth to blaspheme God, and to slander his name and his dwelling place and those who live in heaven. ⁷ <u>It was given power to wage war against God's holy people and to conquer them</u>. And it was given authority over every tribe, people, language and nation. ⁸ All inhabitants of the earth will worship the beast—<u>all whose names have not been written in the Lamb's book of life</u>, the Lamb who was slain from the creation of the world.

⁹ Whoever has ears, let them hear.

¹⁰ "If anyone is to go into captivity,
   into captivity they will go.
If anyone is to be killed with the sword,
   with the sword they will be killed."

<u>This calls for patient endurance and faithfulness on the part of God's people</u>." (Revelation 13:5-10)

Therefore, God's people must stick to the Bible and not modify

God's truth to fit human science—the erroneous science-associated conclusions—because all correct human science-associated conclusions affirm the truth of the Bible.

Genesis 1:1 is God's thesis statement about God's creation of the earth and the universe in the six days of creation. And *Genesis 2:4* is God's summary statement for the same creation events.

Even in this detailed and highly specific record of creation, God was sure to indicate that there are certain things He was leaving out because they are beyond our scope; namely, the creation of God's dwelling place; the creation of the heavenly hosts and the creation of the blob of earth and the water the blob of earth was sitting in at *Genesis 1:2*.

All these were left out of God's creation account in Genesis to show mankind that there are a lot that are beyond mankind's concern. Yet, we do not get it. Because the pioneers of science and astronomy opened our eyes about the much that can be learned from persistent inquiry, the world now believes it could solve all human problems on its own.

Worse still, the world hopes to one day prove that what we call God is nothing but human imagination because of our insecurities. But the world is in for a shocking surprise. The Bible will prevail and all the prophecies of God will come to pass. It is a guarantee from God.

God does not have any interest in justifying anything to anyone. Least among His concerns are those who believe that God and His Bible are archaic fabrications designed to keep mankind in the dark. The word of God is the only authority in anything and His simple word works wonders. For that reason, God has no time to entertain anyone who gets over his head in assumptions and fantasies and is bent on derailing mankind's eternal destiny.

The world did not believe Jesus Christ, not because Jesus was unpersuasive and unpassionate, but because the devil was at work, doing what he does best. On the same token, I would not be surprised by a somewhat disparaging reception to the amazing scientific truth which God has entrusted to me that I am sharing in this book.

I am only a messenger, and my duty is to honorably disseminate the information I was given. But I have no doubt that the message would prove especially useful to those whom it was intended for—those whose hopes would be bolstered by the scientific truths in this book, and my other titles.

There is no doubt that many in the church have abandoned the truth laid out in Genesis Chapter One so that they would be in sync with the latest discoveries of science and astronomy. Unfortunately for them, they are destroying their own faith, and making the grace they received to be in vain *(2 Corinthians 6:1)*.

Science came out of the church and the Bible. The attitude of many of the pioneers of science and astronomy was that God and science has nothing to do with each other, so they kept God out of God's own expert creation. And the church felt safer keeping science separate to avoid diluting the gospel of Jesus Christ; and it has remained that way for centuries. Some of the pioneers of science and astronomy were obstinately pushing the church to modify the texts in the Bible to match their discoveries. The notorious case of the church against Galileo Galilei was one such example.

And once these scientists and astronomers broke free from the church's authority and influence, and partnered with governments, they convinced the world through well-orchestrated government programs and civic efforts, that science is superior to the Bible because the facts of science are here and now, whereas those of the Bible and the church are based on trusting God and obeying His commands, even facing the consequences of disobedience.

Science is knowledge, and all knowledge comes from God. And God created knowledge as a road map to everything in life *(Proverbs 8:12)*.

The pioneers of science were in error in pushing the church to change the Bible to match their scientific discoveries.

Science does not compete with the Bible! Science confirms the truth which God has already perfected, and which the Bible has already authoritatively proclaimed. And science does not lead to faith. Science only confirms what faith has already achieved. That is why the Bible says:

"Now <u>faith is confidence in what we hope for and assurance about what we do not see.</u> ² This is what the ancients were commended for.

³ <u>By faith we understand that the universe was formed at God's command, so that what is seen was not made out of what was visible.</u>"
(Hebrews 11:1-2)

It is not science that confirms it to the believer that the visible universe was formed out of what was invisible. It is faith in the word of God that confirms to the believer that the universe was created out of what was invisible. Believers of God and His Christ had known that for centuries before the Big Bang theory introduced that concept to the rest of the world in 1927.

Any believer, who waits to get the evidence—from science, astronomy or other specialized fields of knowledge—before believing in the things God said in the Bible, simply does not have any faith in God. Faith is simply trusting in the word of God, even if the world's so-called evidence seems to contradict your belief.

And you believe, not because you are stupid, but because you know that the God that tells you so, is reliable and knows what He is talking about. So if you are into science, your God, whose word you have trusted, would then give you the evidence in your work, personally. Your faith in God moves God to unlock mysteries to you—as a personal reward to boost your faith in Him some more.

In essence, faith leads to evidence, but evidence does not lead to faith.

Therefore, when Genesis Chapter One plainly lays it out for you how your God created the earth and the universe, why doubt God? If God does not know; who does? There is nothing that comes out of the mouth of God that is not the absolute truth. He is the truth and there is no other truth!

Mankind makes assumptions and approximations because mankind supposes this and supposes that. But whatever God says is the established truth. Our scientists and astronomers devise intricate and sophisticated mathematics—like the Fourier Series and Taylor Series—just to achieve approximate answers. But your God, my God, uses simple arithmetic to explain the most mystifying truths because He is the one that puts them together to start with.

If you have any faith in God, you must start to trust in all His words, or your faith will erode.

Science must become a compulsory requirement for all Bible scholars if they were to have the knowledge and understanding necessary for recognizing and demystifying the hidden truths of the scientific kind that are strewn all over the Bible. Knowledge is important. And not just religious knowledge—all knowledge! God wants His people to be at the cutting edge of knowledge so that they could keep the world informed of the truth of God as written in the Bible.

Anyone who has knowledge has the ability to quickly spot a deception and raise alarm about it. That is why God exclaimed in the Bible: *"my people are destroyed from lack of knowledge. 'Because you have rejected knowledge, I also reject you as my priests; because you have ignored the law of your God, I also will ignore your children.'"* (Hosea 4:6)

We should all remember that Aaron was not only the high priest. He was also the surgeon general and the chief medical examiner; and had the best medical training of all the ancient Israelites of his time.

As I talked about in greater detail in my other books, and somewhere else in this book, God taught Moses and Aaron how to diagnose and deal with infectious diseases to prevent their spread among the Israelites. Yet God decided not to tell them about the microscopic living organisms that cause these infections because both men and the people of that time did not have enough human advancement and experience to comfortably deal with that fact.

So, God taught them just what they needed to know to effectively deal with the problems. Another four thousand years passed by before God finally revealed the existence of these pathogens to the world. And He did, not through the experts but through a textile trader who was doing his best to advance his trade.

The universe, as majestic and imposing as it is, is a necessary inconvenience to mankind—necessary because God wants mankind to believe in Him without seeing Him; and inconvenient because the universe prevents mankind from seeing God in His majesty, and so be prompted by fear to obey God's every command due to His sheer size and glorious majesty.

So let us look at what the bible tells us will happen when God removes the universe: The world thinks it is impossible for the universe to be removed, but the re-absorption of the universe is a piece of cake for God, and a foregone conclusion according to God's prophecies. God promised that the universe will be rolled back the way it was put on. In essence, God created the universe with a Big Bang (explosion), and He will take out the universe in a Big Bang (implosion).

The Bible says that it took God 24 hours flat to create the universe *(Genesis 1:14-19)*. This is the established truth about the origin of the universe. God created everything at the **"speed of God"** which is infinitely faster than the *"speed of light"* on which all the world's scientific calculations in astronomy are based.

Light has the fastest speed known to man in nature, but light did not create the universe—God did! Therefore the universe was not created at the speed of light. God created the universe at the speed of God. Light is one of the countless creations of the

Almighty God, so light could not compete with God. Nothing can or does.

The Bible also says that the universe will be rolled back very quickly "with a roar" *(2 Peter 3:10)*. Here is the passage from the Bible telling you how quickly and how suddenly God would peel back the universe. Does it sound like a mad man talking? Or does it sound like the mighty God who created it all telling you exactly how He would withdraw what He had put up? You be the judge:

"But the day of the Lord will come like a thief. The heavens will disappear with a roar; the elements will be destroyed by fire, and the earth and everything done in it will be laid bare." (2 Peter 3:10)

God had put the universe in place to conceal Himself from the human beings so that mankind could come to Him through faith and faith alone. Faith is the only way any human being can reciprocate the infinite love God has for him or her. Jesus Christ told Apostle Thomas: *"Because you have seen me, you have believed; <u>blessed are those who have not seen and yet have believed</u>." (John 20:29)*

The universe provided the perfect cover for God to get mankind to "believe when they have not seen," because God knows that there is no human being that has enough courage to disobey Him once they see His real physical size and his glorious majesty.

God is larger than all His creations, put together. He is so large that the entire universe sits on His hand—one hand—that He can clinch His fist over it and snuff out everything in a moment. And that He has us in His hand, so He can keep an eye on each and every one of us, speaks volumes about His love and concern for all mankind.

Therefore, everything the Bible says about God is literally true. God does not embellish His words to intimidate anyone into submitting to His authority. Intimidation is not a mark of love. Intimidation precedes exploitation! If God is interested in intimidating anyone, He would have kept Himself visible to mankind throughout human history.

Nobody tries anything illegal under the watchful eyes of a police officer. That will be very stupid. And such direct law enforcement presence 24/7 will really intimidate even the hardest criminal. The direct presence of God would initiate more fear than law enforcement into every human being; in our current sinful nature.

God's physical presence would have quickly kept everyone in line—the same way the powerful nations of the world use their visibly superior military might to discourage their less powerful opponents from doing anything rash. God is, indeed, larger than anyone could wrap his or her mind around. That is why He is called the infinite God, because He is infinitely large; infinitely powerful; infinitely present; infinitely mighty; infinitely knowledgeable; infinitely wise; and infinitely infinite.

God put the universe in place on Day 4 of creation, after He had allowed His direct light to grace the earth for three days, and after He had set up plant life on the earth in preparation for mankind. God did not leave His throne to build the earth or the universe. By mere word and through His Spirit, God accomplished everything He set out to do. Here is God saying that to you:

""Listen to me, Jacob,
*   Israel, whom I have called:*
I am he;
*   I am the first and I am the last.*
13 My own hand laid the foundations of the earth,
*   and my right hand spread out the heavens;*
<u>when I summon them,</u>
*   <u>they all stand up together.</u>"* *(Isaiah 48:12-13)*

And because creation was all about mankind, God set a day on which He will roll back that barrier that is the universe—which He had put up four days after He had created a place for His beloved mankind. And mankind would finally see Him and live with Him forever.

The earth and the universe are sitting in God's mighty hand. So as soon as God withdraws the universe, God's dwelling place would automatically come into view for everyone on the earth. That was exactly the way it was in the beginning—Day1, Day 2 and Day 3 of creation.

Causing the universe to disappear will be like lifting a much larger ball from a much smaller ball. It will be like moving a superdome away to expose a grain of wheat to direct sunlight. And when that happens, our world becomes one continuum with God's dwelling place; without God moving an inch from His eternal throne. And just like in Day 1 through Day 3 of creation, God's direct light will once again grace all of the earth and everyone therein.

The New Jerusalem which is talked about in the Bible is an extension of the heavenly places (God's dwelling place) and would become a natural extension of our world. When God implodes the universe, which had blocked mankind from seeing God, and our world opens up into God's dwelling place and becomes one with it, all of God's glorious majesty would naturally extend into our current world and transform it to become a new earth. And mankind would come face to face with God. And Christ's promise of going away to prepare a place for us will become a reality *(John 14:2-3)*.

God would not move His throne into the world because the King of kings does not move for anyone. Our world would automatically open up into God's Dwelling place, bringing us right into the heavenly kingdom of God. So now it is easy to see how the New Jerusalem would merge with our current world as soon as the universe is peeled away with a roar, and God reabsorbs all the energy from our current universe just the way He brought them forth in the first place.

On Day 4 of creation, through an unprecedented explosion, God dispersed His energies throughout space—created on Day 2 of creation—to form our universe and block the world's view of His dwelling place; and put His might on display for mankind to see and marvel at. And at the end of it all, through an unprecedented implosion, God would recall to Himself all of that dispersed energies of Day 4 of creation and open up space so that our world and His dwelling place would become a continuum. And with all the energies in the universe reabsorbed by God and space laid bare in all of its expansiveness, our infinitely huge God would come into our view, majestically seated on His throne, with His Son Jesus Christ seated on His right Hand.

Chapter 3

The natural is a progression from the supernatural

Everything about God is real—not magic. The natural is a progression from the supernatural, in much the same way visible light is a part of the Electro-magnetic spectrum! Thus the supernatural becomes increasingly natural to those who obey God and seek His guidance. God set all the natural laws as extensions of the supernatural laws; as in a parallel universe. Everything that applies in nature extends from—and is controlled by—the one that exists in the supernatural.

In much the same way the knowledge of science helps mankind understands the existence and usefulness of the rest of the invisible portions of the Electro-magnetic spectrum; the knowledge of God not only gives believers the understanding of the supernatural but also allows them the use of it as an extension to the natural. For those who know God, the supernatural readily becomes available for use just like the natural is available for use. Jesus Christ said so in *(Matthew 17:20)*.

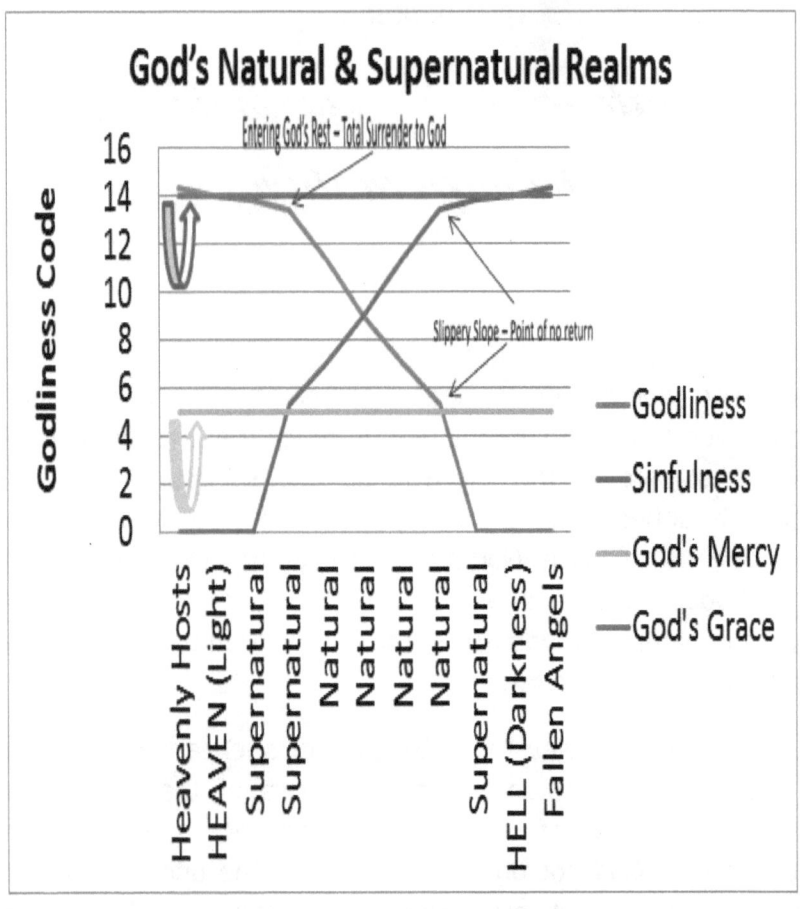

Figure 5: *The Natural is an extension of the Supernatural*

And since the supernatural existence is all spiritual, the spiritual universe governs the physical universe in its entirety. And this is the reason the Bible talks about powers and principalities. Here is the passage for you: *"For our struggle is not against flesh and blood, but against the rulers, against the authorities, against the powers of this dark world and against the spiritual forces of evil in the heavenly realms." (Ephesians 6:12)*

According to this passage, there are spiritual *rulers* with *powers* and *authority* to control this world. These spiritual rulers have their liaisons here in the world with us, who operate out of sight because they are spirits and have authority over us. They interact with us to get us to do as they desire.

And because we cannot see them—most of us do not even believe that they exist. And we cannot initiate any actions to thwart their influence over us. This is why science avoids even the slightest mention of God in anything, because science only deals with things it could capture, dissect and analyze. And the spirits are beyond human control.

The Bible warns that the only way we can escape the vicious webs of these spiritual rulers is by being strong in God and putting on the full armor of God: *"Finally, be strong in the Lord and in his mighty power. Put on the full armor of God, so that you can take your stand against the devil's schemes." (Ephesians 6:10-11)*

"Therefore put on the full armor of God, so that when the day of evil comes, you may be able to stand your ground, and after you have done everything, to stand. 14 Stand firm then, with the belt of truth buckled around your waist, with the breastplate of righteousness in place, 15 and with your feet fitted with the readiness that comes from the gospel of peace. 16 In addition to all this, take up the shield of faith, with which you can extinguish all the flaming arrows of the evil one. 17 Take the helmet of salvation and the sword of the Spirit, which is the word of God." (Ephesians 6:13-17)

One of these dark spiritual rulers of the world was revealed to us in the Book of Daniel as the "Prince of Persia". Here is the passage for you:

"Then he continued, 'Do not be afraid, Daniel. Since the first day that you set your mind to gain understanding and to humble yourself before your God, your words were heard, and I have come in response to them. 13 But the <u>prince of the Persian kingdom resisted me twenty-one days</u>. Then Michael, one of the chief princes, came to help me, because

I was detained there with the king of Persia. [14] Now I have come to explain to you what will happen to your people in the future, for the vision concerns a time yet to come.'" (Daniel 10:12-14)

Jesus Christ also revealed another one of these dark spiritual rulers to us in the New Testament:

"I will not say much more to you, for the prince of this world is coming. He has no hold over me, [31] but he comes so that the world may learn that I love the Father and do exactly what my Father has commanded me." (John 14:30-31)

The two evil spiritual rulers mentioned here are parts of the *spiritual forces* of evil "in the heavenly realms". In the heavenly realms, because their authority comes straight from God—and not because they live in heaven: evil spirits are not rewarded with residence in heaven. Only the angels of God—the good spirits— live in heaven with God.

The evil spiritual rulers are the devil's inner circle, and oversee the activities and operations of the dark forces that are operating within our physical world. All the dark energies in the world, including negative human emotions, sinister schemes and exploitative activities, derive their powers and authority from these dark rulers.

Control of the earth and all of the worldly systems belong to these guys. Our public identification with Jesus Christ and God in our speech and our deeds become troublesome to these guys who try in every way they can to get us back to serving them through our worldly ways.

King Nebuchadnezzar accurately described the powers of heaven in the following passage from the Book of Daniel: *"At the end of that time, I, Nebuchadnezzar, raised my eyes toward heaven, and my sanity was restored. Then I praised the Most High; I honored and glorified him who lives forever.*

His dominion is an eternal dominion;
    his kingdom endures from generation to generation.
35 All the peoples of the earth
    are regarded as nothing.
He does as he pleases
    with the powers of heaven
    and the peoples of the earth.
No one can hold back his hand
    or say to him: "What have you done?"(Daniel 4:34-35)

Look at the phrase, "*the powers of heaven and the peoples of the earth*".
It says it all: All powers are concentrated in heaven from where
they are controlled to affect the "peoples on the earth". Through
the world's governmental systems, mankind believed it has control
over the world. In reality, human governments are simply serving
the supernatural powers which influence the activities and
decisions of the human governments.

That is why the same verse 35 says: *"All the peoples of the*
earth are regarded as nothing," meaning that all the peoples of
the earth have no powers in deciding the overall direction of the
earth and mankind. Look also at the following passage from the
same chapter of Daniel:

"The command to leave the stump of the tree with its roots means that
your kingdom will be restored to you when you acknowledge that
Heaven rules." *(Daniel 4:26)*

And the prophet Isaiah takes it even further, providing us with a
peak into the measure of things, not only as it pertains to power
and authority, but also as it pertains to physical proportions of
God and the various things He had created:

"Who has measured the waters in the hollow of his hand,
    or with the breadth of his hand marked off the heavens?
Who has held the dust of the earth in a basket,
    or weighed the mountains on the scales
    and the hills in a balance?" (Isaiah 40:12)

15 "Surely the nations are like a drop in a bucket;
   they are regarded as dust on the scales;
   he weighs the islands as though they were fine dust.
16 Lebanon is not sufficient for altar fires,
   nor its animals enough for burnt offerings.
17 Before him all the nations are as nothing;
   they are regarded by him as worthless
   and less than nothing." (Isaiah 40:15-17)

23 "He brings princes to naught
   and reduces the rulers of this world to nothing.
24 No sooner are they planted,
   no sooner are they sown,
   no sooner do they take root in the ground,
than he blows on them and they wither,
   and a whirlwind sweeps them away like chaff." (Isaiah 40:23-24)

25 ""To whom will you compare me?
   Or who is my equal?" says the Holy One.
26 Lift up your eyes and look to the heavens:
   Who created all these?
He who brings out the starry host one by one
   and calls forth each of them by name.
Because of his great power and mighty strength,
   not one of them is missing." (Isaiah 40:25-26)

Chapter 4

The Universe completes one full rotation in 365,243 Earth days

What we know from the Bible, Science and Astronomy:

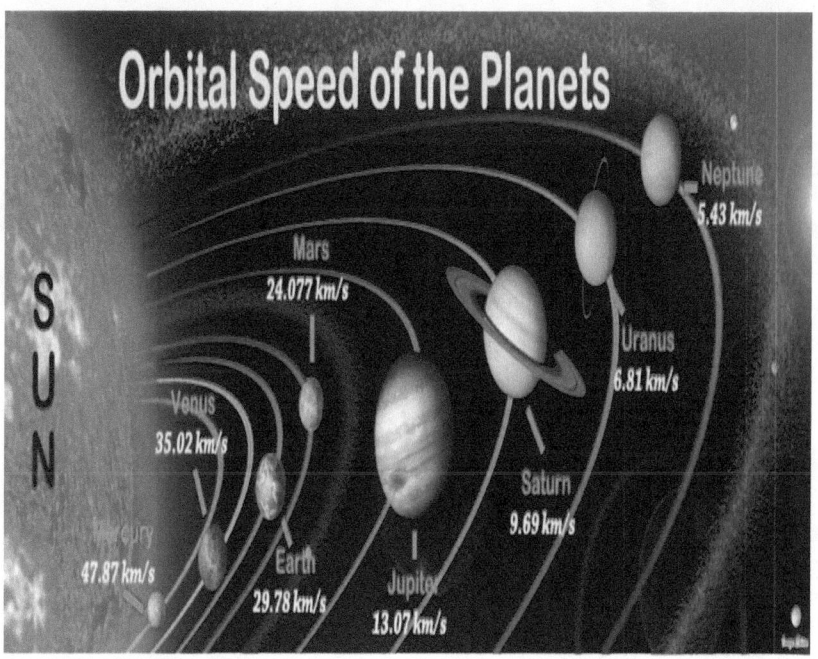

Figure 6: *Orbital Speeds of Planets*

Kepler's 3rd Law				
T in years, a in astronomical units; then $T^2 = a^3$				
Discrepancies are from limited accuracy				
Planet	Period T	Dist. a from the Sun	T^2	a^3	Ratio: a/T
Mercury	0.241	0.387	0.05808	0.05796	1.606
Venus	0.616	0.723	0.37946	0.37793	1.173
Earth	1	1	1	1	1
Mars	1.88	1.524	3.5344	3.5396	0.813
Jupiter	11.9	5.203	141.61	140.85	0.437
Saturn	29.5	9.539	870.25	867.98	0.323
Uranus	84	19.191	7056	7068	0.228
Neptune	165	30.071	27225	27192	0.182
Pluto	248	39.457	61504	61429	0.159

A look at the **"ratio: a/T"**, in the above table, shows that the closer a celestial body is to the source of its orbital energy, the larger the ratio of its astronomical distance to its orbital period. This means that a celestial body closest to the source of orbital energy has more energy available to it, and moves faster around the energy source than a body much farther away from the energy source. See figure 6 above.

Our sun is the smallest known star in the universe and it powers our earth and the other celestial bodies around it. The bigger and more prominent stars in the universe have more energy available for the celestial bodies they power, and the celestial bodies closest to them will move even faster around them than mercury moves around the sun, because of the much higher energy radiating from the larger stars.

For that reason, our universe, sitting in God's powerful hand, would then be boiling over with God's direct radiant energy, making the universe move extraordinarily fast; and causing it to dissipate its excess energies towards its center, to maintain the desired equilibrium within. In other word, God's direct radiant energy sustains the intensity of all the stars inside the universe through this top-down dissipation of energies. See figure 7 below.

We know that the earth and the entire universe—plus everything in them—are completely filled with the Spirit of God. Therefore, God is alive and active throughout the earth and the universe.

Everything God created, including man, exists in God, through His Spirit—God is holy and so nothing convoluted has a place inside God's holy body. God has the universe inside His hand to oversee, control and protect everything He had created, and to keep human mess away from his most holy self. But those who are washed by the blood of His Son Jesus Christ would eventually be given a place in God's direct presence, for their obedience to His commands and their love for their fellow human beings.

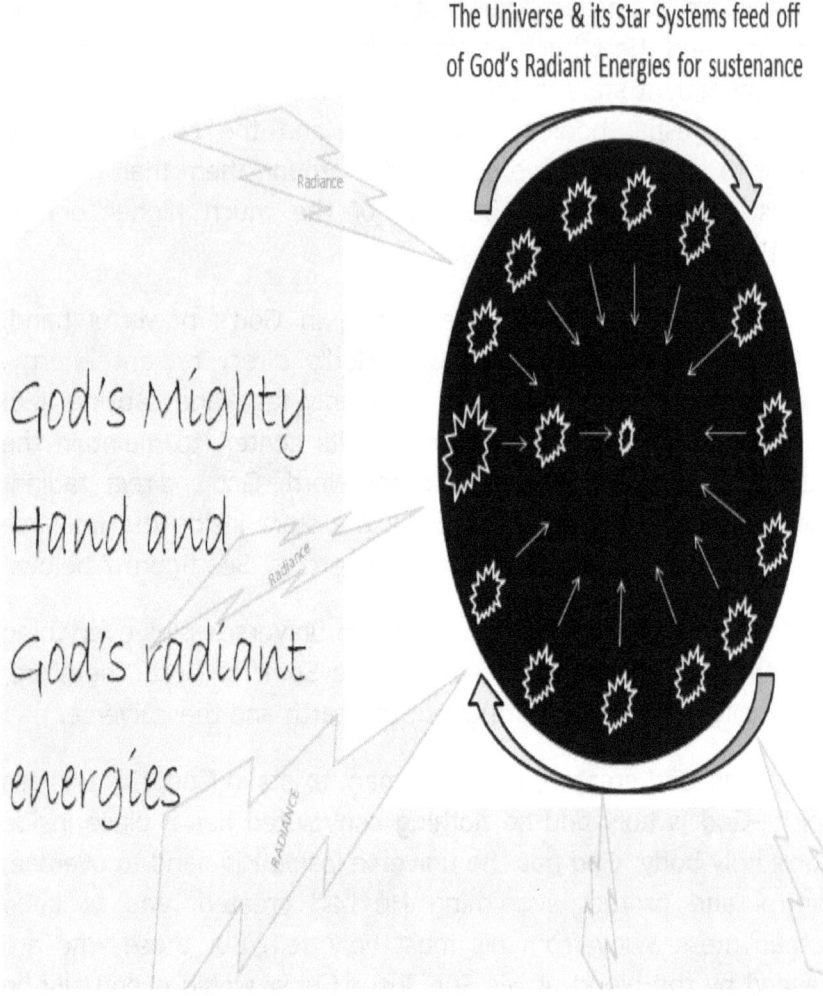

The Universe & its Star Systems feed off of God's Radiant Energies for sustenance

God's Mighty Hand and God's radiant energies

Figure 7: The Universe and its entire star systems continually feed off of God's direct Radiant Energies through a top-down dissipation mode. God has not given human beings the ability to detect this form of energy, nor would He ever do so, because it is the finest spiritual matter in existence.

We know that everything that exists on the earth and in the universe were all created by God, named by Him and catalogued one by one regardless of how small or how large they are *(Isaiah 40:26) (Luke 12:7) among others.*

We know that nothing on the earth or in the universe escapes the eyes of God because God has the entire universe before Him 24/seven *(Isaiah 40:12) (1 Samuel 2:3) (Revelation 5:6) among others.*

We know that nothing could happen anywhere on the earth and in the universe without God first giving His permission *(Job 1:11) (Isaiah 45:7) among others.*

We know that God created the universe in only 48 hours *(Genesis 1:6-8) and (Genesis 1:14-19).*

We know that God makes known the end from the beginning *(Isaiah 46:10).*

We know that everything, without exception, that comes out of the mouth of God happens as God pronounces it *(Isaiah 55:11).*

We know that God's pronouncement of anything means the completion of that thing.

We know that "God is not a man that He should tell a lie".

We know that God gives perception to those who honestly seek the truth and work diligently to get to the truth.

God told us in Genesis that everything He created serves as a sign to something else He created *(Genesis 1:14-19).* And following these signs, we would get to those other hidden things *(Romans 1:20):*

The tiny ants and all their cousins serve as reminders to mankind as to how really small and insignificant mankind is in relation to God. The bacteria and viruses more closely mimic that difference in size between us and God.

Therefore, those who challenge God simply do not understand anything about these signs strewn all around us by God. By giving us the signs, God is making His presence known to all of us in addition to reminding us that it is love that drives Him to put up with our craziness and not because we amount to anything.

The moon's light comes from the sun and the sun's light comes from the original light of *Genesis 1:3-5*; and so does the light from all the other stars in the universe *(Ephesians 4:10)*.

Science talks about exploding stars and how these exploding stars become planets. Yet, no one has observed that transformation, from beginning to finish, in order to demonstrate the validity of the theory. Yet that theory is widely accepted as a scientific fact.

Scientists peer through the telescope and observe two stars, co-joined, and twirling around each other; and declare them merging stars that eventually swallow other nearby stars to form quasi and black holes. But reality on earth has demonstrated to us that twin human babies that are co-joined never merge to form a super human being with double the faculty and facilities of regular human beings.

A half-filled cup does not always mean that the cup started out full. A half-filled cup could have been created that way. That we look into nature—especially far off nature as is the case with the universe—and see things in various configurations does not mean that the configuration are different stages in a natural process of formation.

While that may be the case some of the time, but as it relates to the universe and its mysteries, it is hardly true. The Bible is the approved authority on these matters because the truth in the Bible comes from the only reliable authority that made those things and gave us the facts of His activities and intentions.

All the scientific theories about space and the universe are theories rising from association of the configurations observed in the universe. None of what is theorized in any of these theories has been observed by any human being from beginning to completion. Yet all these theories have been declared facts of science.

All these are nothing but the counterfeit Theory of Evolution at work. Charles Darwin unleashed an untamable beast that could never be restrained. However, all these fell into the prophecies of the end-time in the Bible. Therefore, anyone interested in eternal life—or any life at all—must stay away from all the conjectures and study the truth of God to enter that life.

The human spirit and the human mind represent good and evil, respectively. The spirit in a man gives him life and understanding. When we follow our spirit, we stay in tune with God and on the path of life that pleases God. But when we follow our minds, we go off the path of righteousness, and start to pursue fantasies that lead to destruction.

The earth and the universe represent the permanent and the transient, respectively. The earth and the universe also represent the two sides of the struggle between good and evil in the human life.

While the earth is familiar and yield's clear paths to the important destinations in life, the universe captures one's imagination and inspires fantasies that take one into the uncharted territories where one could easily lose himself.

And also, good is eternal while evil is transient. While good lives into eternity, evil will perish altogether on the Last Day. Both good and evil are real substances on the supernatural side of God's spectrum of natural and supernatural. See figure 5.

Everything God tells us in His gospel is designed and showcased in the human person. The entire human features and psyche constitute the blue print of God's design and construction of the earth and the entire universe. In essence, human biology, psychology and sociology, together, constitute a look into the set-up of the physical world and the physical universe, plus the eternal destiny of man—heaven or hell. And the prophets of God and the disciples of Jesus Christ made proclamations along these parallels.

See this example from Apostle Peter: *"We also have the prophetic message as something completely reliable, and you will do well to pay attention to it, <u>as to a light shining in a dark place, until the day dawns and the morning star rises in your hearts</u>"* (2 Peter 1:21).

{In the underlined part of the passage above, Apostle Peter was alluding to the purpose of the word of God, which stretches the entire length of human existence, from creation to the end of this present universe and time.

God created mankind holy, but by choosing his mind over his spirit, the human person disobeyed God and brought evil on himself. God sent His word, the eternal gospel, to shine as light into the hearts of men (which was darkened by sin).

The gospel is meant to continue to light up this darkened heart of man "until the day dawns and the morning star rises in your hearts"— meaning until Jesus Christ returns to the world and defeats death and darkness; and the universe and its contents rolled away, and the light of God becomes the only light for the saints and the righteous who stayed with the message even in the face of the greatest evil.

God's radiance would then eliminate the need for lamps at night, and the sun for daylight, since nights will be no more. This is the rising of the morning star—and his eternal light—in our hearts.}

Hell included, God designed four concentric circles, superimposed one over another, to represent all of God's created designations— the circles of life! See figure 8. While every human being is designed to end up either in heaven (with God, His Christ and the angels) or in hell (with Satan, the fallen angels and all the evil spirits), there are currently wars on the earth between good and evil;

and wars in the heavens (the universe) between the evil spiritual rulers and the angels of God—unseen to mankind but nevertheless raging on.

Figure 8: The circles of Life!

We also know from the Bible that in everything God has supremacy *(Colossians 1:18-19) (John 10:30).* Here are some samples:

God is the greatest force anywhere (He is the only force on the earth, in the universe and in Heaven).

God is the <u>largest being anywhere</u>, not figuratively but literally (the entire universe fits completely into His mighty hand and with one hand He holds the entire universe up, effortlessly).

God is the tallest.

God is the most elegant and most majestic.

God is the most intelligent being anywhere (The Bible says that only God does wonderful things).

God is the most powerful.

God is the most loving.

God is the most faithful and loyal.

God is the best of everything. All good things come from God. If it is God, He conceived it and then gave it to the world.

The arithmetic of scale shows that the entire universe occupies only a part of God's hand. That is supremacy, size-wise.

We know that all the energies in the universe belong to God and are resident in God. God projects His energy onto/into the things He created and works the energies in them through His Spirit who resides in everything everywhere on the earth, the universe and in Heaven—God's dwelling place.

All the energy sources in the universe were created by God on Day four of creation to take over from God's direct energy that powered the earth from Day 1 to Day 3 of creation; and also to demonstrate to mankind the immensity of God and His endless resources.

We know that God has the universe in His mighty hand, since the water in the hollow of his hand was split into two, one filling the world's oceans and the other segment covering the outer perimeter of space to leave space under vacuum. Here are the passages from the Bible *(Isaiah 40:12) (Genesis 1:6-8 KJV)*.

We also know that God powers the universe from the radiant energies that emanate from Him. *(Hebrews 12:29) (Ephesians 4:10) (Isaiah40:12)*

And we also know that God uses both fire and water to purify anything that is contaminated or needs to be refined. Look at the following passage from the Bible:

"Gold, silver, bronze, iron, tin, lead [23] and anything else that can withstand fire must be put through the fire, and then it will be clean. But it must also be purified with the water of cleansing. And whatever cannot withstand fire must be put through that water." (Numbers 31:22-23)

God's radiance beams against the oceans of fires and the obliterating energies that barricade the outer limits of the universe. And just as the gravitational energies of the sun decreases as you move away from the sun, the radiant energy from these obliterating energies tappers down as it advances towards the center of the universe where our solar system is located.

Even with our sun being very small in comparison to the other stars in the universe, the radiant energies from our sun have to be diverted to other bodies around the sun, to allow just the right amount of energies to be focused on our earth. Otherwise, the fragile balance required to support life on the earth will be disrupted, and all life on the earth will disappear altogether.

Science talks about enthalpies (useful energy) and entropies (dissipated energy) but even the entropies are working energies for God. God does not let anything go to waste. This is why Jesus Christ had His disciples collect all the crumbs from the feeding of the five thousand.

The five loaves of bread and two small fishes did not only feed five thousand people, they also fed birds, wild animals, ants and flies. God has the responsibility to feed all living things and God takes that responsibility very seriously.

So what we call dissipated energy is only un-usable to mankind. But God uses those energies to power other processes of life within the universe, because everything belongs to God, and works together to achieve God's purposes. *(Romans 8:28)*

The universe has to move at an incredible speed since it is receiving its energies directly from God. Anything physically closest to God has to be able to withstand intense heat—and have the right purity—or it is quickly consumed by God's radiant energies. *(Numbers 31:22-23)*

The universe's unimaginable size dictates that only direct energy from God could efficiently power its overall operations.

And since God has the universe in His hand and Jesus Christ His Son is sitting on His right hand, God most likely would not hold the universe away from His only begotten Son, so God likely holds the universe on His right hand so His Son could see whatever God sees.

We know that the universe rotates because the Bible says that it takes the universe 365242.5 earth days to complete one full rotation—one day. *(2 Peter 3:8)*

We know that rotation produces the day and revolution produces the year. That is why one day in planets like Mercury and Venus is longer than their one year, respectively.

We also know that God does not subject Himself to time. God is light so He permanently and continuously shines. That is why before the creation of the earth there was no "day and night"—

only the light of God which flooded and still floods God's dwelling place. The Bible says that God is light and there is no darkness in Him at all. *(1 John 1:5)*

That is why the Book of Revelation says that on the Last day, the universe will be rolled up like a scroll and all the natural stars will disappear *(Revelation 6:14);* and the only source of light will be God Himself. This light will be around 24/seven, making nighttime a thing of the past, forever *(Revelation 22:5).* And all these are literal.

We know that God permanently sits on His throne and is worshipped 24/seven by the heavenly hosts; and that His Son Jesus Christ is seated on His right hand waiting to be sent into the world again to judge all mankind.

Therefore, it is the universe that rotates before God; and not God moving around the universe. God keeps watch over the universe and everything He set up within the universe. And since God has the universe in His hand, the universe does not revolve around God, in the same way the planets inside the universe revolve around their powering stars; because that will be undignified. Nothing gets behind God.

We know that God's dwelling place is situated outside our universe. We know that at the end, the universe will disappear the same way it had come into place: The universe came into existence with a roar and will disappear at the end with a roar.

We also know that everything points to another, and whatever is yet to come has happened before. The world has been filled with the buzz about the Big Bang creating the universe. That discovery did not come out of anything the world has made. God had revealed this mystery of the Big Bang in *Genesis 1:14-19* as the mode by which He created the universe, but the world and the church did not take notice of that.

{A priest of God finally told the world about the Big Bang, but was himself not a big herald of God; but instead, muted the name of God out of God's own revelation—a revelation that gave mankind insight into the work of God, and lend credence to the prophesies of the end, in the Book of Revelation. The priest convinced the Pope and the church to refrain from shouting this gift of God from the mountain top, thereby allowing the secular world the opportunity to secularize the Big Bang, and use it to destroy faith in the God who brought this revelation to the world.

The Big Bang that formed the universe was not the only roar of unprecedented scale in God's creation in the Book of Genesis. The creation of land out of water in Genesis 1:9-10 was another explosion of unprecedented scale; and it preceded the creation of the universe.

The forming of space in Genesis 1:6-8 also created vast, extensive and sustained roar which continued throughout Day 2 of creation. On Day 2, God did not just create space. God separated the water that remains on the earth from the water which He pushed out of the universe as He created space inside the water in the hollow of His hand. And by creating space entirely inside this water, God pulled a vacuum in space as He pushed out the water above space to the outer limits of space, where it still exists today.

Once space was created, God then went on to create orbits throughout space in their different orientations, and powered all of them with their respective speeds and directions of motion. All these were accomplished by God through incredible speed which is infinitely faster than the speed of light, thus permitting God to finish the vast space project within a twenty-four hour period (Genesis 1:6-8).

Here is Apostle Peter telling us that the universe will disappear with a roar just like it came into existence with a roar, **"But the day of the Lord will come like a thief. The heavens will disappear with a roar; the elements will be destroyed by fire, and the earth and everything done in it will be laid bare."** *(2 Peter 3:10)*

And here is Prophet Isaiah telling us about the same end of time because when the sun and the universe implode and disappear, the barrier which

God had erected between His dwelling place and the earth is completely lifted away, and God and the heavenly places instantly become starkly visible to the people that remains on the earth. God's enormous size—the whole universe sits inside his palm—automatically will allow the humans on earth to see God clearly, as one sees another person standing next to him or her. Here is Isaiah's prophecy:

"Every valley shall be raised up,
  every mountain and hill made low;
the rough ground shall become level,
  the rugged places a plain.
⁵ <u>And the glory of the LORD will be revealed,</u>
  <u>and all people will see it together.</u>
For the mouth of the LORD has spoken." *(Isaiah 40:4-5)}*

{The world is spherical in shape with human beings spread across the entire surface of the earth. The universe is also spherical in shape and completely separates us and the earth from God and the heavenly places. When God finally peeled the universe away to reveal Himself and the heavenly places to the world, it would be as though God's dwelling place descended onto the earth. Whereas, it was the world that is opened up to a much more expansive horizon beyond the limits of the universe.

The space immediately above the earth becomes one, unobscured continuity with the space currently occupied by space and the universe. This joined view, in turn, becomes one unobscured continuum with the huge expanse that is God's dwelling place; thereby completing the merger of all three spaces into one seamless and unobscured view that is as vast and expansive as it is glorious and majestic.

The intensity of God's direct light allows God's light to illuminate all of God's dwelling place with light. It is the limited intensity of our sunlight that prevents the sunlight from illuminating the entire spherical earth at once, because of the size of the earth in relation to the size of the sun.

To illustrate, pick up a spherical soccer ball and look at it. Have a second person placed exactly across from you and the ball. At the same time you see your own side of the ball, the second person also sees his

77

own side of the ball. But none of you sees each other's side of the soccer ball because the ball is spherical.

It is the light of the sun that allows you both to see features on the opposite sides of the soccer ball at the same time because the soccer ball is very small relative the size of the sun whose light illuminates the soccer ball's 360 degrees at once.

In the same manner, when God re-absorbs the universe, and the earth is exposed to God's direct radiant energies, the entire earth relative to the size of God will resemble the size ratios between the soccer ball and the sun, thereby allowing the light of God to light up the 360 degrees of the earth at once, permanently.

Even if the earth remains spherical and rugged at the time of the end, the intensity of God's light and the infinite size of our glorious God would allow everything on earth to be flooded with sufficient light, just the same way the sun sufficiently illuminates all sides of the soccer ball when there is light. The earth will sit as an almost invisible tiny spec in the middle of the amalgamated heaven.

The people of the world, without taking a step out of the world, would now come to share a common dwelling with the Almighty God. And the direct light of God, from this point forward, will serve the new united heavenly kingdom continuously and infinitely, thereby completely erasing forever darkness and the evil that lurks in the dark.}

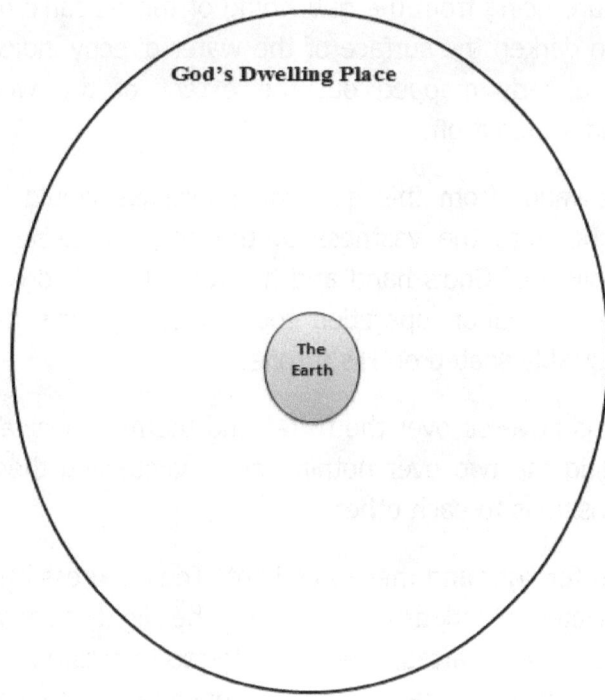

Figure 9: *End-time configuration (Not drawn to scale)*

The Book of Revelation says that God's dwelling place will descend onto the earth on the Last Day so God would dwell with all those who love Him. This is not figuratively speaking. It is literal:

Before God created the earth, the water of creation in *Genesis 1:2* was in the hollow of God's hand; with the earth—a simmering molten sea of volcanic lava—submerged inside the water.

The quenching of this volcanic mass that was sitting inside the water in the hollow of God's hand generated huge amounts of dark dense clouds that enveloped the water and the volcanic mass inside it, creating total darkness on the face of the water.

That the moisture rising from the quenching of this volcanic mass concentrated to darken the surface of the water directly indicates that God has already mapped out the extent of the world's atmosphere and sealed it off.

Otherwise, the vapor from this quenching exercise would have risen even higher into the vastness of the space between the water in the hollow of God's hand and the rest of God's dwelling place: The entire creation operation took place in God's hand, with God comfortably seated on His throne.

The Spirit of God hovered over the water and the mass inside the water suspending the two over nothing and maintaining them in their relative positions to each other

God then called for light and there was light. The darkness on the face of the water receded as soon as the light appeared. Ionization of the gases started, the earth started to rotate so that the light could touch every part of it within each twenty four hour period.

The very instance of the appearance of the light kicked off time. That first instance of the appearance of light on earth is high noon. That is why the Bible says that the evening came and the morning came; and that was the first day. The first day on the earth started at noon and went through the evening, night and into the morning and noontime again, to complete the circle.

On Day 2 of creation, God created space, designed and built all the orbits across space, giving them directions and creating and setting the required energies to propel anything that comes into each orbit. God filled the entire universe with these invisible but extremely fast self-powered highways throughout space in preparation for His creation of Day 4.

And by creating space inside of the water in the hollow of His hand, God "separated waters from waters," pushing the water on top of the space out of space, never to return again into our universe. That water still exists today beyond the limits of our current universe.

And because space was created inside water—the water in the hollow of God's hand—space is dark and cold and eerie. In much the same way the depth of the world's seas and oceans are dark and frigid, our space is dark and frigid because our space is completely surrounded by water of great immensity. This is the water that was pushed out of space when God created space in *Genesis 1:6-8.*

Unfortunately, mankind missed that point because mankind has come to think of God as having size and faculties that are within the vicinity of the ones God put in the human beings. And this is because mankind has failed to follow God and trust His every word as God has commanded mankind. Instead, mankind went off chasing after shadows, trying to elevate itself above God. Eve was only the first; and not the only one. All of humanity has eventually become like her. But now is the time to change, because the end is really at hand.

On Day 3 of creation, God commanded for land to come out of the water in the hollow of God's hand, and an unprecedented volcanic eruption thundered through space, with molten volcano majestically rising out of the still water; forming land that jots out of the water; and triggered humongous tsunamis that raced away from the new land and clapped over the land, cooling the land, weathering it and flooding the low lying portions of the land to form lakes and ponds. Water got trapped within the giant spikes of solidified volcanic rock, forming underground reservoirs that later feed rivers and streams across the land.

Later on Day 3 of creation, God created green vegetation over the land He had brought out of the water and watered through the streams, rivers and standing bodies of water of various sizes. With the earth's atmosphere forming from Day 1 of creation at the appearance of light, the land created and watered, and the sunlight to power photosynthesis, the earth was ready to support life; and it did.

The sealing off of the earth, its vegetation and its atmosphere by the Spirit of God—who was hovering over the water and the earth—protected the earth's installations from the cataclysmic explosion that took place on Day 4 to create the universe.

On Day 4 of creation, God commanded for lights in the expanse of space, triggering the explosion of the original light of Day one through Day 3. Oceans of fires and simmering energies spread across space, and raced to the farthest reaches of space to create the universe.

God pushed the largest stars, quasi and the other oceans of simmering energies much farther away from the earth, leaving only the sun to directly power the earth. God also buffered the earth with other planets to absorb the excess energies from the sun so that just the right amount of sun's energies could end up on the earth.

God also positioned some of these bodies to shield the earth from nuisance celestial bodies that could disturb and even destroy earth's delicate balances that support life on the earth. Nuisance celestial bodies are not necessary but there for the benefit of man—to stir man's curiosity about God and instill fear in him to respect nature and obey God.

God arranged the burning oceans of energies into galaxies, setting their inter-relational energies and movements to add balance to the earth and the life it contains.

God completed His creation of the universe in only 24 hours. But because the world of science has decided to explore the earth and the universe by keeping God out of it, it limited itself to solving a supernatural problem with natural faculty and facilities. Once light was identified as the fastest thing in the universe, science and astronomy decided that they could decipher the mystery of creation by using the speed of light to measure everything in creation.

God wedged the universe between the earth and God's dwelling place—high above the universe where it is all light and serenely majestic all the time—to physically separate the earth and its notoriously ambitious and disloyal occupants from God's chaste and glorious abode. No disingenuous or convoluted thing ever gets close to God or His dwelling place.

We also know the following astronomical information in the table:

	Mean Orbital Distance	Distance from the Sun	One Full Revolution	One Full Rotation
	(Miles)	(Miles)	(Earth's Time)	(Earth's Time)
The Universe	505.8 billion, trillion*	80.5 billion, trillion	N/A	365243 Earth days
Pluto		3.7 billion	248 earth years	6.4 earth days
Neptune	17.5 billion	2.8 billion	164.8 earth years	16 earth hours
Uranus	10.9 billion	1.88 billion	84.01 earth years	17 earth hours
Saturn	5.6 billion	938 million	29.5 earth years	10.25 earth hours
Jupiter	3.1 billion	508 million	11.9 earth years	9.8 earth hours
Mars	898 million	155 million	686.98 earth days	24.6 earth hours
Earth	584 million	93 million	1 year	1 day
Venus		68 million		243 earth days
Mercury		43.5 million	87.97 earth days	176 earth days
Sun's Diameter	228 million			

Notes: The value recorded as the mean orbital distance for the universe is actually the computed diameter of the universe based on science's projection of the edge of the universe being 13.7 billion light years from the earth.

From science's projection that the edge of the universe is 13.7 billion light years from the earth, the rotational speed of the universe computes out to = 1.60273E+13 miles per second (16 trillion miles per second).

If the scientists got it right, the information from the Bible then suggests that the universe is rotating at many times faster than the speed of light. Science could be wrong in its projection of the distance from the earth to the edge of the universe. But if science is correct, that kind of speed for the rotation of the universe is not impossible, because direct radiance from the God of the earth and the universe—whose radiant energies sustains the energies of all the millions of trillions of stars in the universe—could move the universe infinitely faster than mere stars can move the planets surrounding them.

It is, however, important to note that the table above has shown that while orbital speed increases with proximity to the source of orbital energy for a planetary system, rotational speed does not increase with proximity to the energy source—at least as exhibited in our solar system.

Mercury, which is the closest planet to our sun, rotates much slower on its axis than the more distant earth rotates on its axis. While our earth, which is larger than mercury, completes one full rotation in one earth's day, mercury completes one full rotation in 178 earth days; and Venus, which is about the same size as earth, completes one full rotation in 243 earth days.

Here are further notes to consider:

Earth's Diameter = 7,900 miles

Earth's Circumference is 49612 miles

Earth's orbital Distance = 583723738.8 miles

Earth's orbital speed = 29,800 m/s = 66635 mph

Distance from the earth to the Moon 239,000 miles
Distance from Earth to the sun = 92.96 million miles

Earth's main tectonic plates: African plate, Antarctic plate, Indo-Australian Plate, Eurasian Plate, North American Plate, South American Plate, and the Pacific Plate.

The galaxies in the universe are rotating and revolving around the center of universe

The galaxies move in circular paths along fixed radial pathways relative to the center of universe

If the astronomers have accurately estimated the farthest known object in the universe at 13.3 billion light years, then the circumference of the universe is easily calculated from the Bible.

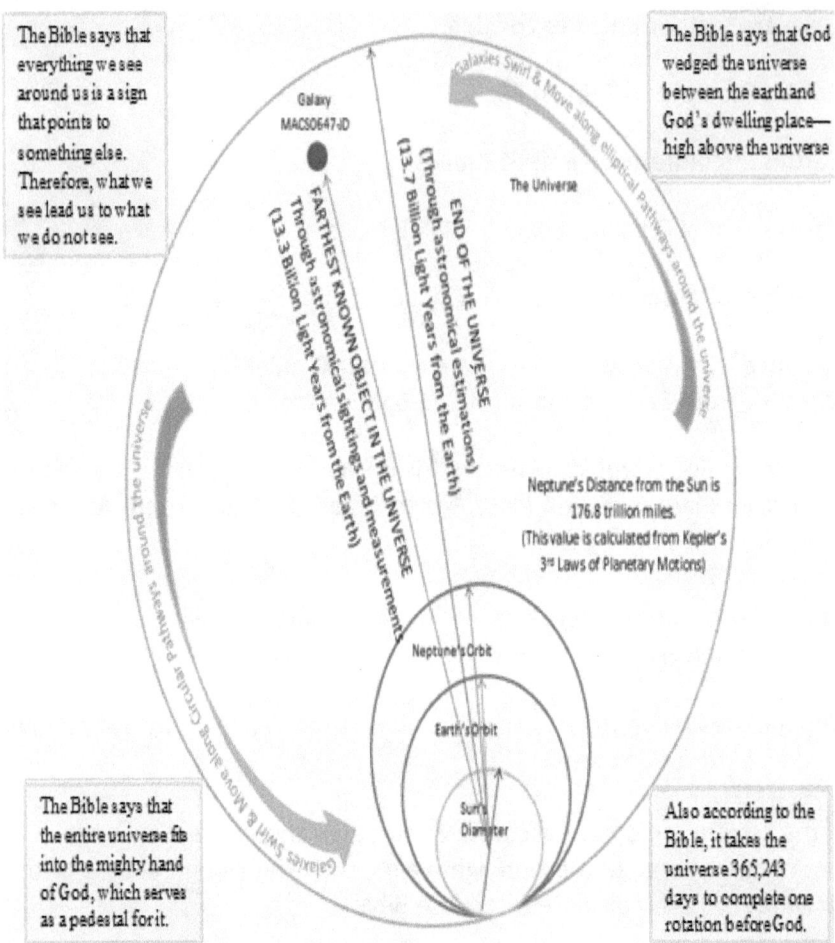

Figure 10: Size of Universe Illustration

Chapter 5

Here is to the Speed of God!

There have been a lot of efforts in the world of science and astronomy to demonstrate to the world that the Bible's account of creation is untrue. Governments around the world have been pouring endless resources into science and astronomy to convince the world that human destiny lies in our own hands; and not in the hands of the God of the Bible. The scientists have been very busy looking all over the universe in search of earthlike planets that shows any form of life at all.

And while all these are going on, they are made to like harmless explorations for the truth. None of these is designed to be harmless. It is all geared towards discounting the truth of God in the Bible. And many of us are saying *"well if it uncovers facts that are contrary to the Bible, so be it. It is better to know the truth than to remain in the darkness in the name of God."*

That is the kind of attitude that destroys utterly. No truth will ever contradict the truth of God in the Bible. But human beings in their dubious schemes could present evidence in ways that may look contradictory to the Bible either willfully or in error.

But whatever the reason may be, such misrepresentation of the truth is still devastating to mankind: it erodes faith among those whose faith in God is shaky, and reinforces the carefree attitude of those who never had any faith to begin with. That makes the world a more dangerous place as people begin to design their own morality and find justifications for their wrongdoings.

Scientists have recently detected a very distant planetary system they believe is "80% as old as the universe". Here is an internet excerpt about this very distant star-planet system called Kepler 444 *(http://en.wikipedia.org/wiki/Kepler-444) :*

Kepler-444 is a star in the constellation Lyra, estimated to be 11.8 billion years old. On 27 January 2015, the Kepler spacecraft is reported to have confirmed the detection of five rocky exoplanets orbiting the star.

The star is approximately 11.8 billion years old , whereas the sun is only 4.6 billion years old. The star is 80% of the age of the universe.[3] It is an orange main sequence star of spectral type KO.[5]

The original research on Kepler-444 was published in The Astrophysical Journal on 27 January 2015 under the title "An ancient extrasolar system with five sub-Earth-size planets"[6] by a team of 40 authors, the abstract reads as follows:

"The chemical composition of stars hosting small exoplanets (with radii less than four Earth radii) appears to be more diverse than that of gas-giant hosts, which tend to be metal-rich. This implies that small, including Earth-size, planets may have readily formed at earlier epochs in the Universe's history when metals were more scarce. We report Kepler spacecraft observations of Kepler-444, a metal-poor Sun-like star from the old population of the Galactic thick disk and the host to a compact system of five transiting planets with sizes between those of Mercury and Venus. We validate this system as a true five-planet system orbiting the target star and provide a detailed characterization of its planetary and orbital parameters based on an analysis of the transit photometry. Kepler-444 is the densest star with detected solar-like oscillations. We use asteroseismology to directly measure a precise age of 11.2+/-1.0 Gyr for the host star, indicating that Kepler-444 formed when the Universe was less than 20% of its current age and making it the oldest known system of terrestrial-size planets. We thus show that Earth-size planets have formed throughout most of the Universe's 13.8-billion-year history, leaving open the possibility for the existence of ancient life in the Galaxy. The age of Kepler-444 not only suggests that thick-disk stars were among the hosts to the first Galactic planets, but may also help to pinpoint the beginning of the era of planet formation."[6]

Planetary system

The Kepler-444 planetary system[4]			
Companion (in order from star)	Orbital period (days)	Inclination	Radius
Kepler-444b	3.60001053	88°	0.036 R$_J$
Kepler-444c	4.5458841	88.2°	0.0443 R$_J$
Kepler-444d	6.189392	88.16°	0.0473 R$_J$
Kepler-444e	7.743493	89.13°	0.0487 R$_J$
Kepler-444f	9.740486	87.96°	0.0661 R$_J$

All five rocky exoplanets (Kepler-444b; Kepler-444c; Kepler-444d; Kepler-444e; Kepler-444f) are confirmed,[4] smaller than the size of Venus and each of the exoplanets completes an orbit around the host star in less than 10 days.[3] The system is also very compact and Kepler-444b is the smallest at 0.403 earth diameters and even the furthest planet, Kepler-444f, still orbits closer to the star than Mercury is to the Sun.

Every object inside the universe is powered by a star, which the object is orbiting. That is why the planets that orbit Kepler 444 have orbital properties comparable to those of the planets that orbit our sun even though Kepler 444 is so close to the outer perimeter of our universe.

The universe, on the other hand, is powered by direct energy that radiates from God and not from some star. In other words, everything inside our universe is a part of our natural world and behaves as all else in nature, but anything outside our universe belong with God's dwelling place which starts from the outer perimeter of our universe. Therefore, God's dwelling place is a physical place that surrounds our universe.

The human science has determined Kepler 444 as being 80% as old as the universe because of its distance from the earth. Their determination was based on the speed of light which is in great error, because God did not create our earth and our universe at the speed of light.

God created our earth and our universe at the "speed of God" which is infinitely faster than the speed of light. This is the error that this book is aggressively trying to bring to light. These men and women of science got it all wrong. Through their accepted theories, they make the following assessment: *"The star is approximately 11.8 billion years old , whereas the sun is only 4.6 billion years old."*

That is a great error. The Bible unapologetically declared that all the stars, the planetary systems and the star systems in the universe, were born on the same instant—*"And God said, Let there be lights in the firmament of the heaven to divide the day from the night; and let them be for signs, and for seasons, and for days, and years: 15 And let them be for lights in the firmament of the heaven to give light upon the earth: **and it was so**." (Genesis 1:14-15)*

Following this command of God to populate space with the trillions of stars and the other celestial bodies in the universe, the Bible says:

*"And God made two great lights; the greater light to rule the day, and the lesser light to rule the night: **he made the stars also**.*

17 And God set them in the firmament of the heaven to give light upon the earth,

18 And to rule over the day and over the night, and to divide the light from the darkness: and God saw that it was good.

19 And the evening and the morning were the fourth day." (Genesis 1:16-19)

God, therefore, spent the next 24 hours of that day—Day 4 of creation—shaping the lights and all the celestial bodies, and setting them in place with their full functions, arrays and inter-relationships.

For this reason, there is no one star in the universe that is billions of years older than any other star in the universe—just like the claim that Kepler 444 star is 11.8 billion years old while our sun is only 4.6 billion years old. Scientists are simply playing a game of numbers, and high-fiving one another for a job well done.

That is absurdity in the face of the truth of the Bible. Scientists are mere men and are rather dabbling into great mysteries they had no clue about. They look at everything in an upside down manner because God has given them over to the deceptions of their own minds.

But the Christians could care less. After all, they have been won over by everything science says. The Bible says that God brought the stars out one by one and called each of them by name, and that because of God's "great power and mighty strength", none of the stars is missing *(Isaiah 40:26)*. And here is that passage for you from the Bible:

*"**Lift up your eyes and look to the heavens: Who created all these?** **He who brings out the starry host one by one and calls forth each of them by name.** Because of his great power and mighty strength, **not one of them is missing**."* *(Isaiah 40:26)*

The Assumptions of Science in determining the Age of the Earth and the Universe:

The following are some of the assumptions that science made in its determination of the age of the earth, the universe and everything in them:

That God has nothing to do with creation. And that time and chance brought the universe into existence. In other words, science believes that everything that exists in the universe has a natural origin and that nature itself is the architect of everything in existence and had created them over long stretches of time.

That the time it takes the light from each star to reach the earth is an indication of how long it took the star to reach its current location from the moment of the Big Bang.

And that since science could predict the distance of objects observed in the universe through some of the properties the observed objects exhibit, the current location of anything observed in the universe directly represents how long ago the object was formed and dispersed through the Big Bang to have reached the location at which it was detected. This is why science projected that the universe was created by time and chance about 13.7 billion years ago.

Science's inherent errors in removing God from creation and everything else:

The "speed of God", at which God created the earth and the universe, is infinitely faster the speed of light. That is why it took God only 24 hours to create the universe in its entirety—in addition to the 24 hours it took God to create space two days earlier.

So by eliminating God from their pursuit of how everything started, they missed the most important factor employed in creation—the speed of God. And to make up for what they have dismissed, they convinced themselves, and the world, that the universe was created long before the earth came into existence, because there is no other way to make sense through their theories.

The universe that took God only twenty four hours to created— and another 24 hours He spent creating space on Day 2 of creation—is now projected as something that took 13.7 billion years to create. Yes! God guarantees us in the Bible that He is doing new things which we know nothing about. Yet, it is God who affirmed that none of the stars he created, and marshalled out one by one, has been lost.

Jesus Christ made the same declaration in relation to the twelve apostles God gave to Him, and we could quickly certify His claim *(John 17:12)*. Yet we remain mute about a similar claim of even a greater consequence which God made about the stars of heaven, and instead allow science to mess with the truth of God. Here is Christ's claim:

"While I was with them, I protected them and kept them safe by that name you gave me. **None has been lost except** *the one doomed to destruction so that Scripture would be fulfilled." (John 17:12)*

This goes to show that either the church is really not paying attention to all the established truths of God in the Bible or that the church has been utterly overpowered and silenced by the world that the church now gullibly accepts everything science produces in the name of the truth.

Christians now look at their God from their human eyes, instead of the eyes of their spirit. It is no secret that, over the centuries, science has steadily won over many Christians and a greater percentage of the Christian leadership, to the point of reducing our great God to the level of a powerful human being; instead of the powerful supernatural God who literally holds the entire universe on one hand—from the creation of the earth and the universe until now. Here are some of the things we need to remember:

God does not compete with anything He created.

God cannot be measured against anything He had created.

God is not within human reach that He should worry about any human being or any other power, period.

God is not a piece of object that could be contained and studied and dissected by lowly human beings.

God does not do anything at the speed of anything He had created. The speed of God is infinitely faster than the speed of anything God had created—light included. How can the created match the creator? If that happens, the creator could be at the mercy of the created in the event of an accident.

God is perfection and does not make a mistake. He gets it right all the time. If He thinks it, it is done and done perfectly.

God is infinitely superior to everything that He had created.

God is infinitely confident about the safety of His secrets from human meddling. He controls all knowledge and decides who He gives knowledge to. Without God, all humanity will disappear altogether.

God is confident that He would realize all His plans without a hitch because all the powers in the world, the entire universe and heaven reside within God. God controls and executes actions for all life—natural and spiritual. Outside of God, no powers or life or energies exist. That is why the Bible says that God does not despise any human being.

God completed the universe in 24 hours flat—not more than that.

The universe was not created before the earth. The earth's starting material was created by God even before our time started in *Genesis 1:3-5*. The earth was submersed in water in *Genesis 1:2*. Then in *Genesis 1:14-19*—Day 4 of creation—God created the universe as a barrier between His dwelling place and the earth so that whoever believes in Him have to do it by faith, and not intimidated by His physical presence.

The scientist do not know it but finding an exoplanet which shows some evidence of life means absolutely nothing, with respect to making their contradiction of the Bible the truth. Lies can never become the truth and the truth can never be subverted. Not in God's world, because our supreme God is alive and well and very much in control of His earth and His universe, including the human lives—with no exception.

It is good that the efforts in science and astronomy are yielding fascinating results. But I will assure everyone, none of that would ever negate God. Everything science has so far uncovered, and everything science would ever uncover, will, one hundred percent, support everything God laid out in the Bible for the believers.

God said that "my people are perishing for lack of knowledge." Scientific knowledge is good for our faith in God and not detrimental to it. I say the things I say in this book and all of my books in confidence because God usually puts the information into my spirit before He leads me to the worldly evidence to support what He already put in my spirit.

I do not start to research scientific information to reach a conclusion. It is the other way around for me. I am usually led to the truth by the Spirit of God. Then in quick order, I am led to the supporting scientific information the world already know. Therefore science only confirms the information that was put in my spirit. And it is my conviction that the Bible is 100% correct, literally.

I believe the Bible completely and will continue to seek the truth; not from science but directly from the God, through the Bible. Science only helps me firm up the things I learn from the Bible. The Bible says, seek and you will find, ask and it will be given to you; and knock and it will be opened for you. It works all the time because God is looking to teach us directly because we are all precious to Him.

Our earth and the universe and everything in them were not created at the speed of light. They were all created at the speed of God which is infinitely faster than the speed of light. That is why it took God 24 hours flat to develop space and set all the orbits and all their properties on Day 2 of creation. That is also why it took God only 24 hours to complete the entire universe on Day 4 of creation.

The farthest galaxy away from the earth and our solar system are exactly the same age as all the other galaxies everywhere in the universe; and also the same age as the universe and everything else within the universe, because God created them all in one 24-hour period. And God assured us in *Isaiah 40:26* that none of the stars are missing because of God's "great power and mighty strength"

According to Genesis Chapter One, God spent 4 out of the six days of creation putting the earth and all of its life-supporting systems together, thereby demonstrating His surpassing love for mankind.

In comparison, God devoted only 2 out of the 6 days of creation in the creation of the entire universe: God spent Day 2 on the creation of space; and Day 4 on the creation of the universe.

Anyone who has problem believing the account of creation in Genesis Chapter One clearly does not understand the power of God. Therefore, the person has a lot of work to do in building his or her faith in God before it becomes too late.

God said in the Book of Isaiah: *"My own hand laid the foundations of the earth, and my right hand spread out the heavens; when I summon them, they all stand up together." (Isaiah 48:13)*

Science revealed that light is the fastest thing on the earth and in the universe. The speed of light is 186,000 miles per second. Now that's fast. But that is because science does not know God. God is the maker of heaven and earth and everything that exists in them, including light. He is the God that called for light in *Genesis 1:3* and instantly there was light, time was born, and the earth went into rotation so that the entire face of the earth could see the light within a twenty four hour period.

Then in *Genesis 1:14-19*, God called for light again, but this time He called for light to be in the expanse of the firmament He had created in *Genesis 1:6-8*—space—that seemingly unlimited expanse that surrounds the earth and its waters. When God created space on Day 2 of creation—*Genesis 1:6-8*—God designed all the orbits in space in their endless configurations and orientations, and set the respective orbital speeds and directions.

And as soon as God commanded for light to fill space *(Genesis 1:14-19)*, through a cataclysmic event, the original light of Day 1 transformed into trillions of islands of fireballs, energies and light and shot through space, propelled by the vacuum God drew when He created space in *Genesis 1:6-8*, reaching the farthest limits of space and filling space completely—all within the 24 hour period that constitutes Day 4 of creation. Now! Here is the speed of God, the fastest speed anywhere.

All the galaxies and their constituent celestial bodies were set in their orbits and set in motion within this time frame. That is why the Bible says, *"He who descended is the very one who ascended higher than all the heavens, in order to fill the whole universe."* *(Ephesians 4:10)*

He is God and there is no other! And apart from Him there is no God *(Isaiah 44:6)!* As soon as a word leaves His mouth, that word has already accomplished what the word is expressing. Some in the church has mistakenly characterized this as the speed of sound, meaning that the Almighty God accomplishes things at the speed of sound since He accomplishes everything through spoken words that generate instant results.

But that impression is scientifically in great error. The speed of sound—768 miles per hour in dry air at 20 degrees Celsius—is only a very small fraction of the speed of light, 186,000 miles per second. But not even the speed of light comes close to describing the speed at which the God of all creation operates. If God works at the speed of sound, God will still be creating the earth, and would not have gotten to the creation of the universe.

Our God does not move at the speed of anything He had created, not even the speed of light (186,000 miles per second). Our God moves at the **Speed of God**. And the speed of God is infinitely faster than the speed of light. Because the universe is so expansive and nobody truly knows the size of it, we still could not fully understand the speed of God unless we clearly examine the Scriptures. And that is what I am going to attempt to give you a sense of how fast the speed of God really is.

When I first encountered what I am sharing with you, I thought it was God being dramatic, even though I have always known that our great God has no time to be dramatic to impress any lowly human being. His Son Jesus Christ expressed it time and time again, and demonstrated that it is real.

We are all very familiar with the claims of science that the universe has been around for billions of years and that our earth has been around for a smidgen less than that. The scientists tell us that the planets, of which the earth is one of, were dead stars that burned themselves out. So, if that were the case, the universe had to precede the earth by billions of years. And all these are because science desperately works to exclude God in the explanation of anything in the world or the universe.

Because the world's measurements have determined that other planets, stars and galaxies are billions of light years away from our earth; the world scientists who assumed that God was not responsible for the Big Bang then believes that by multiplying the distance between the earth and any of these distant galaxies tells us how far ago the event of the Big Bang happened.

The scientists are right if God had nothing to do with the creation of the earth and the universe. But because God is the one who did everything and was sure to tell us that in the Bible, we know that their assumptions are false and their results incorrect by a great margin. In reality, their results are billions of light years away from the truth of God.

The world scientists have missed the greatest concept in the entire world and the universe—the Speed of God. There is nothing like it anywhere. That these distant galaxies and their celestial bodies are billions of miles away from us shows just how fast God operates. The God of heaven and earth fills the whole universe and the earth, and there is no smidgen of them that are unaccounted for. For the God that is that Big who had created us in His own image and likeness, the distance across His mighty hand is the breadth of the entire universe, from end to end.

Therefore, to God, moving anything from one extreme end of the universe to the most extreme end of it on the opposite side simply requires that God slide that thing from one side of His hand to another. And since it will take any of us only a few seconds to complete the described action, it takes God even much less time to accomplish the same feat.

God did not give the human beings better abilities than He gave to Himself! Therefore, moving anything from one end of the universe to the other is **God-fast**—which is infinitely faster that the speed of light.

When the Bible makes declarations relative to God, as in Isaiah Chapter 40, the Bible is not just being metaphorical. The Bible is stating scientific truth about God, and about the things God created in Genesis Chapter One. Through the great prophet Isaiah, God gave the human beings a glimpse into the true size of God by comparing the critical creations of God to the size of God's hand. Nonetheless mankind continues to bring God to its own level; or dismiss Him altogether. Let's take a closer look at that chapter of Isaiah:

Isaiah 40:12 –

- The first line of this verse says that the water in the hollow of God's hand—a cavity or concave area on God's hand—is too deep and too expansive that no one could possibly measure it. This is a reference to the water of *Genesis 1:2* which held the earth at the onset of creation. The earth and the entire water that was present in *Genesis 1:2* were contained within a shallow crack on the palm of God's hand. And this was before God pushed most of the water to the outskirts of the universe as He created space in *Genesis 1:6-8*.

Figure 11: *The Earth, submersed in water in a shallow crack on God's Mighty Hand in Genesis 1:2.*

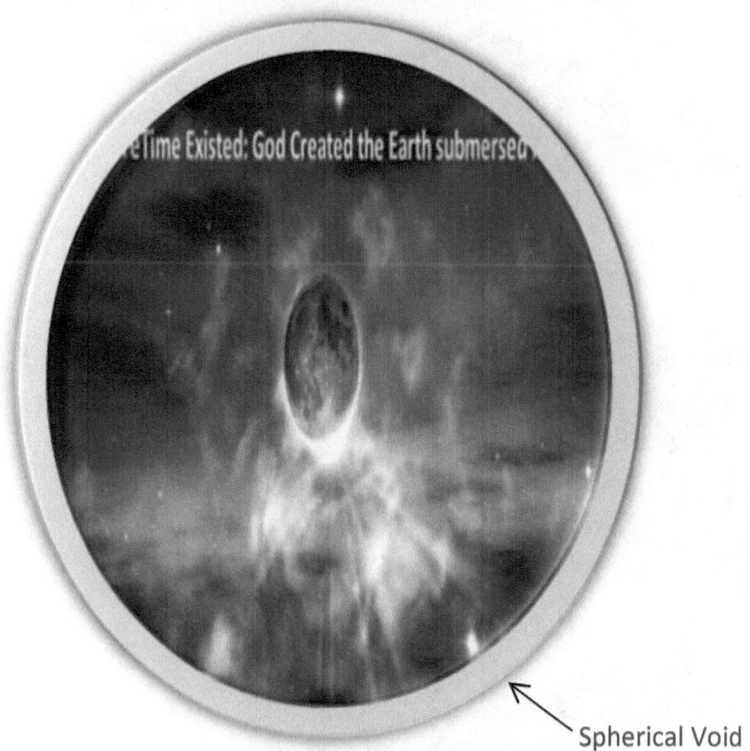

Spherical Void

Figure 12: *God made a spherical void form around the earth, encapsulating the earth and some water which later become the seas on the earth. God, then, radially stretched out the spherical void to create our modern day space— equipped with all the orbits, orbital speeds, orbital configurations and directions. By creating space inside the water in "the Hollow of God's Hand" Isaiah 40:12, God pushed the water on top of space outside our universe where the water still exists today (Genesis 1:6-8); thereby maintaining space under vacuum till the "Big Bang" of Day 4 that created the Universe.*

Figure 13: *God pushed the water on top of the void in Genesis 1:6-8 to the outer perimeter of space. That water never re-entered the universe so it still exist outside the universe till today.*

Figure 14: *On Day 3 of creation, through a cataclysmic volcanic explosion, God created land out of that part of the water that remained on the earth from the creation of space. Huge tsunamis raced across the water over the newly created land and the mountains—* **"the waters stood above the mountains.** [7] **But at your rebuke the waters fled, at the sound of your thunder they took to flight;** [8] **they flowed over the mountains, they went down into the valleys, to the place you assigned for them.** [9] **You set a boundary they cannot cross; never again will they cover the earth"** *(Psalm 104:6-9). The water from the tsunamis cooled the land and weathered it to prepare the land for the planting of green vegetation later on the same Day 3.*

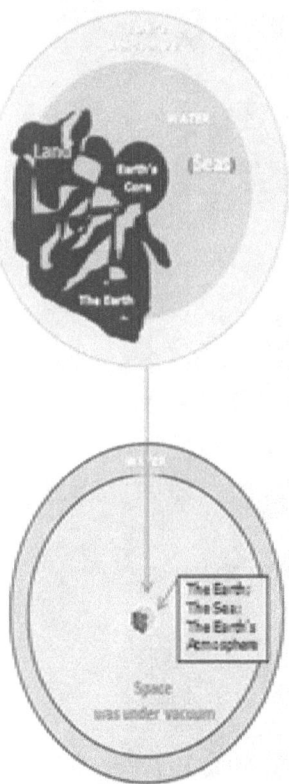

Figure 15: *Land jotting out of the sea after formation; and the sea collecting to one side. Earth's gases were fully formed and sealed off from the adjourning expansive space.*

Figure 16: *On Day 4 of creation, the newly created space reverberated as an unprecedented explosion rocked it and sent oceans of fire flying in all directions of space at the "speed of God"—which is infinitely faster than the speed of light and the universe, in its vast array and infinite configurations, was completed by God in 24 hours.*

- The second line of *Isaiah 40:12* says that the entire universe is the same size as the width of God's hand— ***"with the breadth of his hand marked off the heavens."*** The breadth of the hand (or width) is a distance and not a marking tool. This sentence is saying that with the span of God's hand, God measured out the heavens—which is the universe. And if the universe is only the size of God's hand, God is indeed outside of the universe because the rest of God could not fit into God's hand.

And if the universe is only the size of God's hand, imaging how huge the rest of God really is. And since the universe in its entirety fits into God's hand, moving anything from one end of the universe to the farthest end of the universe is actually much faster than moving the thing from God's left hand to His right hand.

Figure 17: The fully formed universe stays suspended over nothing in God's mighty hand *(Isaiah 40:12)*; and rotates before God. One full rotation of the universe in God's hand makes one "universe day". And one "universe day" is equivalent to 365,243 earth days *(2 Peter 3:8)*.

- And the last line of *Isaiah 40:12* says that God *"__held the dust of the earth in a basket__,"* and *"__weighed the mountains on the scales and the hills in a balance.__"* God is giving humanity a comparative description of the earth and its mountains and hills. To us, humans, these things are of huge sizes. But to God who is infinitely bigger than anything He created; and the earth being very tiny in comparison to God's other creations, the earth could be put into a little basket; and its mountains and its hills could simply be weighed on scales and balances. In reality, God does not need any instruments either to create or to measure. God is all in all; and knows everything with precision!

Isaiah 40:15 –

- *"__Surely__ the __nations are__ like __a drop in a bucket__; they are regarded as dust on the scales;""* The Bible is saying here that all the nations of the world put together is equivalent to a drop of water into a bucket. Isn't that what science has determined of the size of the earth as compared to the rest of the universe? The whole earth is the size of a tiny drop in a bucket that is the universe.

  And the Bible makes this immensely accurate comparison long before any human being acquired the ability to project knowledge in scientific realism. And from *Isaiah 40:12*, we see that the universe—which is referred to here as the bucket, fits perfectly into the hand of God.

  All the nations put together constitute the whole earth; all the people on the earth, and everything they own—including all the tapped and untapped resources within the nations' geographical boundaries and political and financial systems.

- *"He weighs the islands as though they were fine dust."* God reckons the accurate size of the land masses on the earth which are separated by bodies of water—small and large—as though these land masses are the size of fine dust in comparison to the much larger masses and vastness He had also created.

This, once again, is an accurate comparison in respect to the bucket that is the universe, the earth that is a drop in the bucket, and now the earth's land masses which are equated here to fine dusts.

Figure 18: *The earth's land masses are referred to here, in Isaiah 40:15 as islands, and collectively regarded as fine dust in relation to the universe and God's other creations of massive sizes.*

Isaiah 40:16 –

- *"Lebanon is not sufficient for altar fires, nor its animals enough for burnt offerings."* The Bible is saying here that the entire area that constituted the ancient Lebanon with all of its dense and towering forests and famed timber is very small compared to God's fireplace; and all the animals living in Lebanon's forests are small in comparison to God's real burnt offerings. This verse is designed to give God's people who sometimes give to God grudgingly that their offerings amount to nothing in God's reality. If the nations of the earth, taken together, amount to less than nothing; the size of Lebanon, then, is much less than nothing, in true size comparison.

Isaiah 40:17 –

- *"Before him all the nations are as nothing; they are regarded by him as worthless and less than nothing."* The Bible is saying that all the nations—that is the entire earth and all the people of the earth, their possessions and the earth's natural reserves—amount to "nothing."

This means that collectively, they measure up to nothing in the grand scheme of things. The people, their world communities and all their wealth and knowledge put together, do not even begin to show up on the scale with which God measures everything. Therefore, collectively, the world communities are so insignificant in comparison to everything God created that they can easily be discounted on the ground of their true size and material importance alone.

God puts everything in this demonstrative scale to show mankind that the only reason mankind matters is because God wanted mankind to matter; and not because mankind amounts to anything. God had created mankind in His own image and His likeness to groom mankind as His own children and cut us in on Jesus Christ's inheritance, so we could live eternally with Christ and God and the angelic hosts; and share with Christ forever.

And since it was God who created the earth, and on the fourth day of creation, created the universe, getting these distant celestial bodies as far away as they are from the earth did not take anywhere as much time as the scientists thought it did. God moved these distant bodies away from the earth at **God-speed, which is infinitely faster than the speed of light**, allowing God to transverse the entire universe within a twenty-four hour period or less, as the Bible stated in Genesis 1:14-19. God created the universe in only twenty four hours—and not more.

The creation of the universe—the billions of galaxies and their trillions of trillions of stars, quasi, and all the other celestial bodies—might actually have taken God less than twenty four hours to finish. This translates into just a small fraction of a second for God to propel the farthest celestial body away from the earth from the moment of the Big Bang, set it in cluster with others celestial bodies, build gravities around each of them, set them in orbital motions and inter-relational motions, set a purpose for them, name them and document His work and His accomplishment of them; and moved on to the next cluster, until He finished work on all of them as He had set out to accomplish.

I give it to you then that the farthest celestial body from the earth took God a miniscule fraction of one second to move it from the site of the Big Bang to its current location which is billions of light years away from the earth. This concludes then that God-speed is infinitely faster than the speed of light which is 186,000 miles per second.

It establishes that what takes light billions of years to transverse, took God only a miniscule fraction of a second to transverse. Now that is impressive! Let's all give God a praise.

Science says that the light that reaches us today from a distant star that is, say, a million light years away, left the star a million years ago to be able to reach us today. That is correct if God has nothing to do with anything. The fact that the star that is a million years away from us was taken from somewhere around the earth and moved to its current position by God suggests that the light we see today from that star is not a million years old; but rather a part of a streak that the light from the star left on its pathway as it was moved from where it was created to the distance of one million miles from the earth.

To illustrate my point, have someone shine the bright light from a powerful flash light into your face from a short distance from you. Then, have that person run backwards away from you, holding that light still pointing to your face.

As the distance between you and the receding flashlight increases, you will continue to see the light from the flashlight. If the person holding this flashlight moves so far away from you that the person and the light gets out of your sight because of the great distance between the two of you—even though the person is still in a straight line from you and the ground between you and the person remains flat—you will no longer see the light, because the face of the flash light is not large enough to remain visible from that distance.

But if you acquire a telescope that can see far enough, because the person and the flashlight is still on your line of sight, you will once again pick up the light from the flashlight; not because the light from the flashlight just travelled to you, but simply because the telescope minimized the distance between you and the light and brought the light in focus again.

Conversely, if the person holding the flashlight somehow increases the circumference of the flashlight and the intensity of the light as they moved away from you, the light will continue to stay in focus as the distance between you and the flashlight increases.

That is what is going on with the lights we see in the sky directly with our naked eyes or through powerful telescopes. We pick up light from anything that produces light because the source of light is large enough to be seen, and the intensity of the light is strong enough to reach us; not necessarily because of how long the light has to travel to get to us.

In other words, when we see light from anything, it is our sight that travelled to the light to join with it; and not the light from that source that travelled to our eyes to announce its arrival to our eyes. We see all lights through pathways that were established between our eyes and the source of the light. And these pathways are well-established continuum of light that follow the trail of the light source.

In our current scientific belief, the light that left a star a million light years ago retains all of its properties through the million years that the light travelled from that distant star until it gets to our eyes, not counting all the planetary and stellar motions everywhere on the pathway of the light, among the billions of other light pathways originating from other light sources and travelling in all directions of the universe.

If that is true, then the reverse is equally true: the light that a distant star left behind, as it was moved by God to its current very distant location—which is a million light years away from the earth—is still around and completes the continuum of light which we see when our eyes pick up the distant star that is associated with the light. The pathway of light that connects from our eyes to the distant star did not originate from the distant star's current distant location, but rather originated from around the earth and left a trail of light that allows us, even today, to see that distant star.

That being the case, a star that is a million light years from the earth has not been around for a million years because we can see its light on earth today. A million light years simply describe how far away God moved that star from the earth when God dispersed the stars throughout the universe.

From the record of creation in Genesis Chapter One, and other places in the Bible, we learn that it took God only seconds to move the most distant star to its current location. And God moved these stars that far out to give mankind a sense of how great and how fast God is.

Regardless of whether anyone in this world is a scientist or a trash collector; a believer of Jesus Christ or someone without a religion; or anything else in this world; the Bible assures us:

"__For no one can lay any foundation other than the one already laid, which is Jesus Christ.__ [12] If anyone builds on this foundation using gold, silver, costly stones, wood, hay or straw, [13] their work will be shown for what it is, because the Day will bring it to light. It will be revealed with fire, and the fire will test the quality of each person's work. [14] If what has been built survives, the builder will receive a reward. [15] If it is burned up, the builder will suffer loss but yet will be saved—even though only as one escaping through the flames.

[16] Don't you know that you yourselves are God's temple and that God's Spirit dwells in your midst? [17] If anyone destroys God's temple, God will destroy that person; for God's temple is sacred, and you together are that temple.

[18] Do not deceive yourselves. If any of you think you are wise by the standards of this age, you should become "fools" so that you may become wise. [19] For the wisdom of this world is foolishness in God's sight. As it is written: "He catches the wise in their craftiness"; [20] and again, "The Lord knows that the thoughts of the wise are futile." [21] So then, no more boasting about human leaders! All things are yours, [22] whether Paul or Apollos or Cephas or the world or life or death or the present or the future—all are yours, [23] and you are of Christ, and Christ is of God." (I Corinthians 3:11-23)

Therefore, whatever anyone decides to do in this life is their choosing and there are consequences for every choice we make. If in our study of science we choose to chase after fallacies and in the process deceive the world, there is a price to pay for that.

We should all look closely to this passage and see where we belong in it and adjust accordingly so we would not cut ourselves off from the kingdom of God because we are bent on selling our fantasies and lies to the world.

God has laid the foundation for all mankind and each and every one of us is expected to build off of that foundation. Doing it any other way puts us in a collision course with destiny. And we'll only have ourselves to blame.

Just as no one, but us, know the thoughts in our respective minds; only the Spirit of God knows the thoughts of God *(1 Corinthians 2:11)*.

It, then, becomes imperative to know that only the Spirit of God brings any human being to acknowledge God and be able to proudly proclaim that "Jesus is Lord" *(1 Corinthians 12:3)*.

Therefore, anyone who could not bring themselves to accept the truth of the Bible and acknowledge the lordship of Jesus Christ should be concerned, and should ask God to show them the right path to life.

It is God's speed that made the earth and the universe the enigma which they are today. Any human being who has discounted the speed of God has denied Himself or herself the understanding of what makes the world and the universe work.

The speed of God is the reason why Jesus, after His resurrection and His encounter with Mary Magdalene, was able to go to His Father in heaven and came back to the earth the same evening; showed Himself to many disciples, and met with His twelve apostles as He had told Mary that morning.

And the speed of God is how a simple prayer brings God's intervention at a split second to those who call on Him and trust in His might—the power that transformed the earth and the universe into the wonders that we know today.

But because the world scientists are thinking that the earth and the milky way separated from everything else from the moment of the Big Bang at which everything went its own way; and that these celestial bodies travelled to their current locations being propelled by the unaided force of that original blast.

So by multiplying the distance from these very distant bodies to our planet by the speed of light they obtained what they believed was how far ago the Big Bang took place—in other words, the age of the universe. And from that vantage point, they deduced the age of everything else that exists in the universe.

What they completely missed by taking God out of their theories and calculations, is that they missed the speed of God which is infinitely faster than the speed of light. And their punishment was to look at everything backwards to make any sense at all out of things.

In other words, when they removed the real wonder from creation, they ended up with the speed of light. And the speed of light is like snail-pace in comparison to the speed that set everything in the universe in place—the speed of God. And because it was the speed of God that set everything in place in the universe, what was set a billion light years away from the earth, in reality took God a split of a second to be moved that far away from the earth.

Although the Bible clearly says that God first created the earth, even before He dawned light on the earth, science refuses to acknowledge God in any way, shape or form, in spite of His ubiquitous presence; and the fingerprints He left on everything.

The world had thought they could carry on without God. But now God has revealed that the world has gone too far in the wrong direction because of its pride and shortsightedness.

In departing from God, the world had done all its calculations in reference to light and has erroneously concluded that the earth and the universe are billions of years older than they really are. And with that conclusion came many other illusions that are far from the truth of God. That a celestial body is located a billion light years away from the earth does not mean that that body has been around for a billion years, because it only took God seconds to move it that far away from the earth as He distributed these bodies across space to create the universe. Here again is the Bible telling you how long it took God to create the entire universe:

"And God said, Let there be lights in the firmament of the heaven to divide the day from the night; and let them be for signs, and for seasons, and for days, and years:

15 And let them be for lights in the firmament of the heaven to give light upon the earth: and it was so.

16 And God made two great lights; the greater light to rule the day, and the lesser light to rule the night: he made the stars also.

17 And God set them in the firmament of the heaven to give light upon the earth,

18 And to rule over the day and over the night, and to divide the light from the darkness: and God saw that it was good.

19 And the evening and the morning were the fourth day." (Genesis 1:14-19)

And here is Jesus Christ communicating the same thing God wrote down for us in Genesis Chapter One. In His interaction with the Samaritan woman at the well, Jesus Christ said to the woman: *""Woman," Jesus replied, "believe me, a time is coming when you will worship the Father neither on this mountain nor in Jerusalem. [22] You Samaritans worship what you do not know; we worship what we do know, for salvation is from the Jews. [23] Yet a time is coming and has now come when the true worshipers will worship the Father in the Spirit and in truth, for they are the kind of worshipers the Father seeks. [24] God is spirit, and his worshipers must worship in the Spirit and in truth."" (John 4:21-24)*

Watch the secret unfold here. In verse 23 of the above passage, Jesus told the woman that "*a time is coming and has now come.*" This is not a pattern of speech. The God of heaven and earth, who accomplishes so much with so little, does not have time for embellishments. Jesus Christ simply pronounces things as they occur—as He receives them from God the Father.

Jesus Christ announced the coming of a time when everyone who worships God must worship Him in truth and in Spirit, and not care about where they worship Him. And immediately after He made this announcement of the time that was coming, He concluded that the time he had just talked about had just arrived as He was still talking—hence *a time is coming and has now come.*

This is the speed of God. It happens so fast that you can barely catch it. If something is coming, it is coming from somewhere and requires time to get to wherever it is going. But anything that comes from God gets from God to its destination in an instant, because God fills the whole earth and the entire universe to the brim with His Spirit. Therefore God moves things from one end of Himself to another end of Himself in less than the time it takes the fastest among us to move anything from one hand to another hand.

Our second example is the boldest demonstration of the speed of God by Jesus Christ. Jesus Christ showed Himself to Mary Magdalene in the morning of his resurrection, saying to her, "*Do not hold on to me, for I have not yet ascended to the Father. Go instead to my brothers and tell them, 'I am ascending to my Father and your*

Father, to my God and your God.'" (John 20:17).

That was on the morning of His resurrection. Then later the same Day, Jesus met with His disciples as He had told Mary; clearly demonstrating that in a matter of a few hours, He had gone to the Father in heaven; had counsel with the Father; and returned back to the earth; to take care of His disciples, as planned.

In yet another example, after Jesus Christ told the Jews that where he was going they could not follow, He repeated the same statement to His apostles:

""My children, I will be with you only a little longer. You will look for me, and just as I told the Jews, so I tell you now: <u>Where I am going, you cannot come.</u>

³⁴ "A new command I give you: Love one another. As I have loved you, so you must love one another. ³⁵ By this everyone will know that you are my disciples, if you love one another."

³⁶ Simon Peter asked him, "<u>Lord, where are you going?</u>"

Jesus replied, "<u>Where I am going, you cannot follow now, but you will follow later.</u>"" (John 13:33-36)

In verse 33, Jesus Christ categorically told the apostles that they could not come to where he was going. And from that point to verse 36 could not have taken more than a couple of minutes. Yet, that was enough time for God to move.

In verse 36, He said that they could come to where He was going, only later. Jesus Christ does not speak half-truths, nor does He correct Himself. His word is as good as gold anytime it leaves His mouth. From *"you cannot come"* to *"you cannot follow now, but you will follow later"* are worlds apart. Between those two spoken truths supernatural events transpired to lead from one to another. That is the speed of God.

It may not look that way to a world intellectual, but the spiritual man will give thanks to God for His fulfillment of a promise—where Christ is, His servants will also be. And in a matter of minutes, God made that a reality, not just for the twelve, but also for all His disciples to the end of time.

And here is our next example of the speed of God pertaining to Jesus Christ. This time He was talking to His apostles and His apostles who demanded that He told them how to get to where He was going as He had just announced. They also asked that He showed them the Father so they would be content. Here is that passage from the Bible:

""Do not let your hearts be troubled. You believe in God; believe also in me. ² My Father's house has many rooms; if that were not so, would I have told you that I am going there to prepare a place for you? ³ And if I go and prepare a place for you, I will come back and take you to be with me that you also may be where I am. ⁴ <u>You know the way to the place where I am going.</u>"

⁵ Thomas said to him, "Lord, we don't know where you are going, so how can we know the way?"

⁶ Jesus answered, "<u>I am the way and the truth and the life. No one comes to the Father except through me. ⁷ If you really know me, you will know my Father as well. From now on, you do know him and have seen him.</u>"

⁸ Philip said, "Lord, show us the Father and that will be enough for us."

⁹ Jesus answered: "<u>Don't you know me, Philip, even after I have been among you such a long time?</u> Anyone who has seen me has seen the Father. How can you say, 'Show us the Father'? ¹⁰ Don't you believe that I am in the Father, and that the Father is in me? The words I say to you I do not speak on my own authority. Rather, it is the Father, living in me, who is doing his work. ¹¹ Believe me when I say that I am in the Father and the

Father is in me; or at least believe on the evidence of the works themselves. [12] *Very truly I tell you, whoever believes in me will do the works I have been doing, and they will do even greater things than these, because I am going to the Father.* [13] *And I will do whatever you ask in my name, so that the Father may be glorified in the Son.* [14] *You may ask me for anything in my name, and I will do it (John 14:1-14)*

By leading off with the comment, *"You know the way to the place where I am going,"* in verse 4, Jesus Christ provoked the thoughts of His apostles, prompting Thomas to object to knowing the way as Christ had indicated. Jesus followed it up in verse 6 with, *"I am the way and the truth and the life."* And instantly, the apostles saw the way to where Jesus was going and they had no more questions regarding how to get to where Jesus was going.

In the same manner, Jesus again in verse 7 got His apostles thinking about what the Father really looks like by stating, *"If you really know me, you will know my Father as well. From now on, you do know him and have seen him."* And this time around, it was Philip who asked that Jesus showed them the Father so they could know Him and be satisfied. But Jesus came back with *"Don't you know me, Philip, even after I have been among you such a long time?"*

And once again, Jesus Christ led His apostles to see the Father as He had pointed out to them earlier. They all saw the Father through their spirit's eyes: The Spirit of God gave perception to the eyes of their spirits, and instantly their desires were fulfilled. They believed and were satisfied and had no further questions.

Chapter 6

The Bible is the Foundation of Science & of all Knowledge

Come with me! Come with me to the Bible—the ancient manuscripts dictated and authored by the divine Creator of everything that exists on the earth and in the whole universe and beyond. Yes beyond! For His work is not limited to the places man can see or man's mind could reach. His creations go beyond the limits of the universe as man has been able to contemplate.

His creations are limitless and would stupefy anyone who dares to gaze into the scope of it or endeavor to contain it with words. His mysteries are endless and His girth fills the whole earth and the entire universe completely and overfills them into infinity. Yet God remembers it all and every detail He had ever created. And this is in real time. There are no exaggerations here. This is a picture of the God who created you and I, and everything else in the universe and beyond.

God gave us the Bible and hopes that we will follow His guides in obedience and faithfulness so He would uncover for us some of the mysteries He had tucked away in that Holy Book. The mysteries in the Bible are endless and they are all scientific and real. The Bible is not only the foundation of science and all knowledge; the Bible is the fountain of super-science and a window into God's beautiful and infinite mind. Blessed are those who believe and are rewarded by His knowledge.

And only humility, through faith and obedience to God's commands could get anyone close to God's love and enlightenment. God enjoys being our teacher like we humans enjoy being in a position to impart knowledge on our children. Come to the Bible and God will teach you. God does not discriminate because all mankind belongs to Him.

Do you have a scientific mind and an inquisitive spirit? Come to God and validate what you know in science thus far; and then expand in scope and understanding about the things you do not know in science yet. God has it all. Whatever your specialty is in science, He would lead you in the direction you cannot believe is possible.

You will never operate in frustrations when you allow God to teach you science. And remember, when God gives, He inundates you with that which He is giving to you. Try God and learn from His Bible! In the Bible, a simple truth is a subject by itself. Volumes could be written out of one simple fact which God had put in the Bible for the benefit of those who love Him and hunger for His wisdom.

The Bible is the foundation of science. Everything mankind knows so far in science has not even begun to scratch the surface of the great scientific facts God tucked away in the Bible. Because of lack of faith in the one God who made everything, human science has taken a great detour into misinformation, instead of enlightenment. God had promised to confuse the minds of men if they refuse to come to the light and be enlightened and He had done just that.

In 1927, God decided to allow men into His mystery on how He created the universe. He opened that window a little to a Belgian priest in the name of Georges Lemaître, and his theory of the Big Bang set the world on fire, and refocused scientists on a path that had generated tremendous progress in science and astronomy. But instead of men finding God through this window and finally giving God the glory He deserves for all the beautiful work He has done, science took a position of defiance and set out to discredit

God altogether. The Theory of Evolution picked up steam and quickly was position as anti-God and antichrist.

Even Georges Lemaître, the priest that proposed the Big Bang theory, quickly detached the theory from religion and God, even showing frustration at Pope Pius XII for citing his theory as a validation for Catholicism. The priest chose to credit his knowledge of astronomy and physics for his discovery, instead of crediting God's divine inspiration. Here is an internet excerpt concerning Lemaitre's ambiguity:

"By 1951, Pope Pius XII declared that Lemaître's theory provided a scientific validation for Catholicism. <u>However, Lemaître resented the Pope's proclamation, stating that the theory was neutral and there was neither a connection nor a contradiction between his religion and his theory</u>. When Lemaître and Daniel O'Connell, the Pope's science advisor, tried to persuade the Pope not to mention Creationism publicly anymore, the Pope agreed. He persuaded the Pope to stop making proclamations about cosmology. While a devout Roman Catholic, <u>he was against mixing science with religion</u>, though he also was of the opinion that these both fields of human experience were not in conflict." (http://en.wikipedia.org/wiki/Georges_Lema%C3%AEtre)

One of the "Notes and References" to the above internet excerpt says the following about the Belgian priest:

"Simon Singh (2010). Big Bang. HarperCollins UK. p. 362. ISBN <u>9780007375509</u>. It was Lemaître's firm belief that scientific endeavour should stand isolated from the religious realm. With specific regard to his Big Bang theory, he commented: 'As far as I can see, such a theory remains entirely outside any metaphysical or religious question.' Lemaître had always been careful to keep <u>his parallel careers in cosmology and theology on separate tracks, in the belief that one led him to a clearer comprehension of the material world, while the other led to a greater understanding of the spiritual realm</u>... ...Not surprisingly, he was frustrated and annoyed by the Pope's deliberate

mixing of theology and cosmology. One student who saw Lemaître upon his return from hearing the Pope's address to the Academy recalled him 'storming into class...his usual jocularity entirely missing'". (http://en.wikipedia.org/wiki/Georges_Lema%C3%AEtre)

A priest of God whom God had chosen to allow insight into a great mystery had instead chosen to ally himself with the populist world instead of allying himself with the great God of the earth and the universe who allowed him into this great mystery. At his request, and to help save him the embarrassment of ridicule at the hands of his fellow secular physicists and astronomers, even the Pope succumbed to this priest's pressure not to tie the great revelation he had just received from God, with God. What a missed opportunity!

And by commandeering a revelation from God and turning it into something that it is not, the blessing that came from God was quickly turned against God: the Big Bang theory became a physical version of the Theory of Evolution, which was proposed by Charles Darwin half a century prior. So now, with the biological world and the physical universe pointing in the same general direction, the world scientists quickly concluded that it was not necessary to evoke God in explaining the origin of anything in the world or the universe. They begin to openly attack religion and the Bible and ridicule those who bring up the name of God.

God had given this revelation to the priest because God was ready to let the world into the mystery He hid in plain sight in Genesis Chapter One and other places within the Bible, but the priest was too scared of the world to credit this revelation to God. If the world saw this man's contribution to science as a revelation, the world might not give him the credit he deserved and his claims might be quickly dismissed by the world's scientists, so Lemaitre maintained strides with his colleagues that science and religion do not mix.

Even Albert Einstein, who was on record as saying that not all mathematics lead to correct theories, because Lemaitre's math was correct, shortly dropped his opposition to Lemaitre's Big Bang Theory and sided with Lemaitre. Here is that excerpt from the same internet excerpt above:

"Both Friedmann and Lemaître proposed relativistic cosmologies featuring an expanding universe. However, Lemaître was the first to propose that the expansion explains the redshift of galaxies. He further concluded that an initial "creation-like" event must have occurred. In the 1980s, Alan Guth and Andrei Linde modified this theory by adding to it a period of inflation.

Einstein at first dismissed Friedmann, and then (privately) Lemaître, out of hand, saying that not all mathematics lead to correct theories. After Hubble's discovery was published, Einstein quickly and publicly endorsed Lemaître's theory, helping both the theory and its proposer get fast recognition." (http://en.wikipedia.org/wiki/Georges_Lema%C3%AEtre)

Before finally accepting Lemaitre's theory, this was Einstein's reported position: *"At this time, Einstein, while not taking exception to the mathematics of Lemaître's theory, refused to accept the idea of an expanding universe; Lemaître recalled him commenting "Vos calculs sont corrects, mais votre physique est abominable" ("Your calculations are correct, but your physics is atrocious.")"*

Therefore, it is easy to see that mathematics may be correct and the theory the mathematics points to may be incorrect. These great names in science and mathematics all knew it but did not say that much in the public. If someone as famous for his mind as Einstein says it, he sure had seen instances of that being the case. But a world that would stop at nothing in its attempt to discredit God and confuse the people would quietly and conveniently sweep this comment under the rug and continue with its propaganda. That a priest of God had his first priority as being accepted and recognized by the world ahead of his vocation and commitment to God is more than troubling, but he was not the first and had not been the last to choose recognition by the world over association with God.

Mr. Lemaitre was quick to point out to the world that his mathematics and physics led to his theory. If his theological knowledge and background lent to him making the discovery—as I suspect it did—Mr. Lemaitre was sure to conceal that fact from the world so he could receive the world's accolade; like his secular colleagues in science.

Lemaitre promptly shut the Pope down because he did not want the Pope messing it up for him by infusing theology into his theory. And the Pope obliged because either way, the church would still savor the credit for contributing players in such esteemed secular field of study. After all, it was the church that started it all, and still lends brainpower to science.

By proposing that the universe was created from a Big Bang and that the universe continues to expand and accelerate with time, Mr. Lemaitre knowingly or unknowingly opened the door to scientists believing that even though God caused the Big Bang, that the universe's expansion and other activities within the universe had carried on without any need for God. And this is evident in the following comment about Mr. Lemaitre:

"It was Lemaître's firm belief that scientific endeavour should stand isolated from the religious realm. With specific regard to his Big Bang theory, he commented: 'As far as I can see, such a theory remains entirely outside any metaphysical or religious question.' Lemaître had always been careful to keep his parallel careers in cosmology and theology on separate tracks, in the belief that one led him to a clearer comprehension of the material world, while the other led to a greater understanding of the spiritual realm... ...Not surprisingly, he was frustrated and annoyed by the Pope's deliberate mixing of theology and cosmology. (http://en.wikipedia.org/wiki/Georges_Lema%C3%AEtre)

By making his theory a mere physical phenomenon, Mr. Lemaitre overlooked the most critical aspect of the revelation—the speed of God, and God Himself. This is a disservice to God and an unnecessary challenge to God's own account of creation in the Bible. What was intended to share more light on God's immense greatness, ended up being derailed into a use that takes God out of creation.

By eliminating the speed of God—which is infinitely faster than the speed of light—from the theory of creation, Mr. Lemaitre and the world's scientists, substituted time and the speed of light as the most essential criteria for the universe being what it is today; when in reality it was the speed of God and the genius of God that accomplished everything in creation. Thus, a revelation that was supposed to illuminate the truth on the pages of the Bible was turned around to make caricatures of God's record of creation in the Bible, and turned religion on its head.

The speed of God created the earth and the universe and everything in them in six days flat. The entire universe was created by God in 48 hours flat—*Genesis 1:6-9* and *Genesis 1:14-19*—Day four of creation. But the world scientists had proposed that the universe took billions of years to build. And that the earth resulted from a burnt out star; and is much younger than the universe.

This is contrary to what the Bible recorded: the earth was created by God prior to the start of time—*Genesis 1:2*. And that land came out of the water of *Genesis 1:2* through a huge cataclysmic volcanic eruption on Day 3 of creation, with green vegetation planted by God on the same day.

Then with the Spirit of God protecting the fragile earth and its atmosphere and the plant life, God caused the light that dawned on the earth on Day 1 to cataclysmically transform into trillions of oceans of fire that raced at the speed of God through space to populate the universe. The expanding universe and the orbits and their associated speeds and directions were not a result of the Big Bang but rather a design of space created by God on Day 2 of creation *(Genesis 1:6-8)*.

Even science agrees that God has to be much faster than the light He created. Otherwise, He would not have been able to impart that much speed on light. But because scientists are so ready to dismiss the existence of God, they are relying on light as the fastest thing in existence. And even the Christians do not understand the basis of their calculations in projecting the age of the universe.

The scientist's point of reference is fallacious and as such throws all their calculations off by a factor that is in billions. And once God is removed from their theories about how everything started, it made sense that the universe had to precede the earth. And this once again is a lie according to God's own record which is the Bible. Once you remove God from anything, you throw yourself off track and science has done exactly that. It is their loss, not God's.

The Bible is not only the foundation of science and of all knowledge. The Bible is a living book in which every human life, through the entirety of human existence, is recorded. At any given time in anybody's lifetime, one can find himself or herself in the Bible—not only where one currently is in life, but also where one had been and where one is going.

The Bible is a gateway into the Book of Life which is mentioned many times in the Bible. Every human life is captured like a video in the Bible regardless of religion or lack thereof. And peering into the Bible allows one to see where one is and what one must do to get to where God wants one to get to. This is why becoming intimate with the Bible allows individuals to walk with God and His Christ. Do not miss that opportunity. It is a gift from God to all mankind.

Chapter 7

Our Universe is a Catalogue of Signs from God

The Bible says: *"And God said, Let there be lights in the firmament of the heaven to divide the day from the night; <u>and let them be for signs,</u> and for seasons, and for days, and years:*

[15] And let them be for lights in the firmament of the heaven to give light upon the earth: and it was so.

[16] And God made two great lights; the greater light to rule the day, and the lesser light to rule the night: he made the stars also.

[17] And God set them in the firmament of the heaven to give light upon the earth,

[18] And to rule over the day and over the night, and to divide the light from the darkness: and God saw that it was good.

[19] And the evening and the morning were the fourth day." (Genesis 1:14-19).

In Genesis Chapter 1, verses 14 through 19, God created the universe and laid out the starry hosts for the following purposes:

1. *To divide the day from the night;*
2. *for signs,*
3. *for seasons,*
4. *for days,*
5. *For years. (Genesis 1:14).*

The fields of science and astronomy have taught us so much

129

about how the sun divides day from night (sunset to sunrise); how it creates the seasons (back and forth from Tropic of Cancer to Tropic of Capricorn); how it concludes the day (24 hour rotational period); and how it creates the years (365.25 days revolution period).

Science and astronomy has also taught us about the natural signs controlled by physical laws of nature: Moisture rises from the earth's surface and forms a cloud, concentrates and condenses and returns to the earth as rain. They taught us how differences in barometric pressures result in wind movements and wind directions, and their effects on hurricanes and tornados.

But neither science nor astronomy has taught us anything about the supernatural signs—the really ominous ones. All supernatural signs follow laws too, but laws of a different kind. Only faith in God provides anyone the power to discern these signs. Everyone sees them but only those to whom the power has been given can understand them. And these are really the most important signs for which God created the universe for mankind. The universe was created for mankind and not mankind for the universe.

As it is, the understanding of a great percentage of Christian teachers have been diluted, and in some cases rendered ineffectual, by the proliferation of secular knowledge which has forcefully brought questions to the accuracy of the Bible's account of things.

The need for coexistence, and the financial and economic realities of life, does not make it easier but rather puts pressures on Christians everywhere to tone down the Bible message to make it more palatable to people reeling from this uncertainty about the claims of the Bible.

It has come down to Christians preachers expectantly looking for anyone who can tie God or His claims in the Bible to science and astronomy. If that is the only way we would convince anyone out there to believe in the Bible, we definitely have lost the power that comes with the gospel. God wants people to hear and believe. He is not interested in anybody who waits for the scientific proof before he can believe. That is no faith at all, the Bible says: *"For in this hope we were saved. But hope that is seen is no hope at all. Who hopes for what they already have?"* (Romans 8:24)

The Bible says: *"So then faith cometh by hearing, and hearing by the word of God."* (Roman 10:17). All anybody needs in order to believe in God is the word of God! Anybody who needs more than the word of God to believe God, does not have faith at all, and would fall away as soon as the next body of scientific evidence comes in.

Any Christian or preacher who wants to be effective as a disciple of Christ has to have confidence in the simple word of God. We do not convert anyone. We present the gospel and the Holy Spirit does the rest. Overwhelming scientific evidence should never be a tipping point for anyone to believe in God and His Christ; rather it should add to one's joy that he had not believed in vain. But not to help him believe!

God created the universe to serve mankind by being a source of signs to mankind, in addition to the other purposes mentioned in *Genesis 1:14*. And anyone who looks at anything in the universe, and sees that as pointing away from God, that person is in trouble: the devil is playing dangerously with the person's mind.

No one should feed any such urge; but rather, should urgently fight it and return to the path of life and salvation—what you do for a job notwithstanding. If the universe inspires anyone away from God, because the person's job has to do with researching and reporting things in the universe, that person is gaining the world and losing his soul.

And there is nothing more tragic than that. Applying the truth of God to your career will open more doors to you; not close doors. But destroying your faith in God in the name of a career is not wise. God's people do excel in science. So science is not an impediment to anyone having a strong faith in God. It is how one approaches a career in science that could be detrimental to faith and salvation.

A secular astronomer peers through the telescope and sees distant stars merging with one another. He scratches his head, thinking about the natural physical laws that best explains the phenomenon. Whereas a discerning man—vocation not withstanding—makes the same observation or hears news about the same, and immediately one or two Scripture passages comes into his mind. Very quickly, He comprehends what was taking place and in his heart praises God for His remarkable loyalty.

The secular scientist receives human acclaim explaining away the observed phenomenon through the laws of Physics because in his belief everything in the universe has only a physical explanation so he labors dutifully to find answers for his audience so he may be rewarded for his services. And when he could not explain it, he offers a new hypothesis and promises to keep slugging away until he finds an answer.

On the other hand, a discerning man has no theories and no hypothesis. He has only his spirit and God to lean on. He always gets it right but has no credentials to stake his claims on and look credulous. But he knows what is going on with great assurance and is not perturbed by all the commotions and frenzy surrounding anything because He trusts in His God who is always faithful.

But because he is charged by faith with the responsibility of sharing the truth with the world, he says what he knows but nobody believes him; or is even paying attention to him. But he is not disturbed by their unbelief because he has discharged his duties as he was supposed to.

And the fact remains: there is nothing under the sun that is not designed, engineered and brought to completion by the God of heavens and the earth. And that is with no exception. Every idea that a human being has come up with anywhere in the world, and at any time in human history, has come directly from God. Man is incapable of coming up with an original idea.

The 'dark ages' is all God. The ancient is all God. The jet age is all God and the future will also be all God. There is virtually nothing that happens outside of God. And to be sure, anything outside of God simply is not really in existence—it is a figure on someone's imagination. Our all-powerful God makes sure of that.

Any human concept that is valuable which humanity takes credit for inventing is actually a gift from God. God puts an idea into a man and guides the man to its usefulness so that the man would help put the concept into common human use—or even a tragic use, as the case may be. God assures us in the Bible:

"I form the light and create darkness,
 I bring prosperity and create disaster;
 I, the LORD, do all these things." (Isaiah 45:7).

""See, it is I who created the blacksmith
 who fans the coals into flame
 and forges a weapon fit for its work.
And it is I who have created the destroyer to wreak havoc;
17   no weapon forged against you will prevail,
 and you will refute every tongue that accuses you.
This is the heritage of the servants of the LORD,
 and this is their vindication from me,"
declares the LORD." (Isaiah 54:16-17).

Astronomers and scientist race to explain everything they observe in the universe and the distant galaxies because they believe that the universe came to be by chance; when in fact it is the God of the earth and the universe doing His work. Even Christians have this erroneous idea that God stopped working after the sixth day of Genesis Chapter one. That is far from the truth. Check out the following passages from the Bible:

"In his defense Jesus said to them, "My Father is <u>always at his work to this very day</u>, and I too am working."" *(John 5:17).*

In six days, in Genesis Chapter One, God created the '*unsearchable'* earth and the expansive universe. Since, according to Jesus Christ in the above passage, God continues to work, imagine how much God has accomplished since He finished creating the earth and the universe! Read the following passage from the Book of Isaiah and here God confirm this to us:

"Listen to this, you descendants of Jacob,
*   you who are called by the name of Israel*
*   and come from the line of Judah,*
you who take oaths in the name of the LORD
*   and invoke the God of Israel—*
*   but not in truth or righteousness—*
² you who call yourselves citizens of the holy city
*   and claim to rely on the God of Israel—*
*   the LORD Almighty is his name:*
³ I foretold the former things long ago,
*   my mouth announced them and I made them known;*
*   then suddenly I acted, and they came to pass.*
⁴ For I knew how stubborn you were;
*   your neck muscles were iron,*
*   your forehead was bronze.*
⁵ Therefore I told you these things long ago;
*   before they happened I announced them to you*
so that you could not say,
*   'My images brought them about;*
*   my wooden image and metal god ordained them.'*
⁶ You have heard these things; look at them all.
*   Will you not admit them?*

<u>"From now on I will tell you of new things,</u>
<u>   of hidden things unknown to you.</u>
<u>⁷ They are created now, and not long ago;</u>
<u>   you have not heard of them before today.</u>

So you cannot say,
  'Yes, I knew of them.'
[8] You have neither heard nor understood;
  from of old your ears have not been open.
Well do I know how treacherous you are;
  you were called a rebel from birth.
[9] For my own name's sake I delay my wrath;
  for the sake of my praise I hold it back from you,
  so as not to destroy you completely.
[10] See, I have refined you, though not as silver;
  I have tested you in the furnace of affliction.
[11] For my own sake, for my own sake, I do this.
  How can I let myself be defamed?
  I will not yield my glory to another.

Everything we see on the earth and in the universe points to something else the Lord has in store for mankind. All of God's creation is a chain of signs, each pointing to another in almost an unbroken link because God is leading mankind to find Him and desire His love.

By reflecting the sun's light, the moon points to the reflection of the true light—Jesus Christ—by the sun, the stars and the other energy sources.

The Black hole points to the parallel physical and supernatural worlds—immediate absence from the physical world means immediate presence in the supernatural realm. *"For to me, to live is Christ and to die is gain. [22] If I am to go on living in the body, this will mean fruitful labor for me. Yet what shall I choose? I do not know! [23] I am torn between the two: I desire to depart and be with Christ, which is better by far; [24] but it is more necessary for you that I remain in the body. [25] Convinced of this, I know that I will remain, and I will continue with all of you for your progress and joy in the faith, [26] so that through my being with you again your boasting in Christ Jesus will abound on account of me."* (Philippians 1:21).

The physical world is symbolized by the tumultuous side of the black hole while the supernatural world is symbolized by the opposite side of the black hole. Everything that enters the black hole is first condensed to zero mass and exits onto the other side in a new reconstituted form.

Passage through the black hole is instantaneous and goes only from the tumultuous side to the tranquil side. But the Spirit of God goes both ways: from the tumultuous natural side to the serene supernatural side; and from the serene supernatural side back to the natural world. Here is Jesus Christ making that movement, both ways, shortly after His resurrection:

Jesus Christ said to Mary Magdalene ""*Do not hold on to me, for I have not yet ascended to the Father. Go instead to my brothers and tell them, 'I am ascending to my Father and your Father, to my God and your God.'"* (John 20:17).

"*On the evening of that first day of the week, when the disciples were together, with the doors locked for fear of the Jewish leaders, Jesus came and stood among them and said, "Peace be with you!" [20] After he said this, he showed them his hands and side. The disciples were overjoyed when they saw the Lord. (John 20:19-20).*

[21] *Again Jesus said, "Peace be with you! As the Father has sent me, I am sending you." [22] And with that he breathed on them and said, "Receive the Holy Spirit. [23] If you forgive anyone's sins, their sins are forgiven; if you do not forgive them, they are not forgiven." (John 20:21-23).*

[24] *Now Thomas (also known as Didymus), one of the Twelve, was not with the disciples when Jesus came. [25] So the other disciples told him, "We have seen the Lord!" But he said to them, "Unless I see the nail marks in his hands and put my finger where the nails were, and put my hand into his side, I will not believe." (John 20:24-25).*

[26] *A week later his disciples were in the house again, and Thomas was with them. Though the doors were locked, Jesus came and stood among them and said, "Peace be with you!" [27] Then he said to Thomas, "Put your finger here; see my hands. Reach out your hand and put it into my side. Stop doubting and believe." (John 20:26-27).*

²⁸ Thomas said to him, "My Lord and my God!" (John 20:28).

²⁹ Then Jesus told him, "Because you have seen me, you have believed; blessed are those who have not seen and yet have believed."(John 20:29).

Jesus Christ said to Mary Magdalene not to touch him because He *"have not yet ascended to the Father" (John 20:17)* on the first day of His resurrection. Later the same day, He appeared to His disciples and breathed on them to impart the Holy Spirit on them.

It becomes clear that between the time Jesus told Mary not to touch Him and the time He visited His disciples that Jesus Christ had finally "***ascended to the Father***" and came back to the earth for this visit with His disciples. And this is only a space of hours to ascend to His Father in heaven; have consul with the Father, and return to the earth, so He could visit His disciples and interact more closely with them.

From His command to Mary Magdalene that morning, it was clear that Jesus intended to go to His Father first before He could meet with His disciples: *"Jesus said, "Do not hold on to me, for I have not yet ascended to the Father. <u>Go instead to my brothers and tell them, 'I am ascending to my Father and your Father, to my God and your God.</u>'" (John 20:17).*

Christ never said one thing, and did another. His message to His disciples through Mary was *"<u>I am</u> ascending to my Father and your Father, to my God and your God."* The tone of His message *"I am ascending"*—not *"I will be ascending"*—clearly indicates that His ascending to the Father in heaven was imminent and could not wait another second.

Therefore Jesus did ascend to God the Father in heaven before returning to the earth to visit His disciples later that day—*"on the evening of the first day of the week."*

That Jesus Christ hung around after His resurrection and gave Mary Magdalene the message for His disciple was evidence that Jesus was setting something up for the world to notice—the relationship between the natural realm and the supernatural realm and how God owns both and masterfully covers both.

For anyone who is still in doubt that Jesus ascended to His Father, our Father, in heaven from the time He spoke to Mary Magdalene to the time He visited His disciples on the evening of the same day, here is another scenario to consider. Jesus visited His disciples again seven days later and made Thomas swallowed his words.

Between that first day and the time Jesus returned and confronted Thomas— *"Put your finger here; see my hands. Reach out your hand and put it into my side. Stop doubting and believe."* *(John 20:26-27)*— is only a space of seven days. Therefore, from *"Do not hold on to me ..."* *(John 20:17)* to *"Put your finger here; see my hands. Reach out your hand and put it into my side ..."* *(John 20:26-27)*, there are only seven days.

It then becomes an irrefutable scientific fact that it took Jesus Christ between a few hours and up to seven days, to ascend to God the Father in heaven—beyond the outer limits of our universe—have counsel with God the Father, amid jubilation of the heavenly hosts; and return to the earth to visit His apostles and impart the Holy Spirit on them.

In reality, however, Jesus Christ did accomplish that round trip in a few hours within that first day of His resurrection. By faith, I know that He did. And it is that faith that brought me into the understanding of the scientific truth which Jesus Christ established in this passage of the Bible.

It is important to remember that whatever God can do, God

allows His Son to also do, because the Son is the spitting image of the Father and through the Son, the father in which all of God's fullness dwells.

And if we look further, we can see that Jesus Christ could have easily ascended to God and returned back to the earth in a matter of seconds—not even hours: Take the case of the two disciples Jesus joined on their way to Damascus who were talking about His crucifixion; the two who invited Him for supper and a sleep-over to prevent Him from having a run-in with criminals on His way. Jesus accepted their offer of hospitality and went with them.

The resurrected Jesus Christ sat down with them to eat. Once He started to offer thanks over the meal, the disciples realized it was Him and reacted in joy. And <u>instantaneously</u>, Jesus Christ left them.

God does not play games. If Christ had to leave like He did, His departure from the physical world must have meant His appearance in another real dimension—the supernatural dimension; and not an illusionary place. Illusions are for human beings, not for God!

This is the black-hole effect. There is no transitory stop-over somewhere inside the black hole. Whatever goes in must instantaneously come out of the other end in a different existence. Therefore, the black hole is yet another sign of God to lead mankind to discover the hidden work of God that relates to our own life and eternity.

And wherever Jesus went as He left that table, the Father was there too because He and the Father are one. So gone from here in a flash had to mean present in heaven in a flash. In all the instances recorded in the Bible, as soon as Christ left this physical world, He was home with God. God's design is perfect and God gives Himself only perfection.

This is why Apostle Paul declared *"For to me,* **to live is Christ and to die is gain.** *²² If I am to go on living in the body, this will mean fruitful labor for me. Yet what shall I choose? I do not know! ²³ I am torn between the two: I desire to depart and be with Christ, which is better by far; ²⁴ but it is more necessary for you that I remain in the body. ²⁵ Convinced of this, I know that I will remain, and I will continue with all of you for your progress and joy in the faith, ²⁶ so that through my being with you again your boasting in Christ Jesus will abound on account of me."* *(Philippians 1:21)*.

The Bible declared that Jesus rose to heaven and is seated on the right hand of the Father. So if away from life means present with Jesus Christ in heaven, there is really no stop-over anywhere. There are only two forms of existence: here in nature, or in some supernatural realm—heaven for those who love God, His Christ and fellow men and women; or hell for the haters of God who love selfishness, pervert justice and incite violence.

Apostle Paul, through his faith in God and his selfless service to humanity, knew his next destination—present with Jesus Christ in heaven with God the Father. Apostle Paul and all the other disciples of Jesus sacrificed their lives to make it easier for us to choose the same thing they chose for themselves—life of sacrifice on the earth so we may have eternal life with God and His Christ in heaven.

And for those who might argue that Jesus Christ's ascension to heaven in full view of His disciples, forty days after His resurrection, did not happen in a flash; the speed of His ascension was designed for the benefit of His beloved disciples. If He ascended in a flash, the disciples would not know exactly which way He went. But once He went out of site, the speed of God took over as always. Everything Jesus Christ did for the benefit of His disciples was well orchestrated to reinforce their trust in Him and God the father.

Therefore, the ascension was gradual so that there would be no doubts in the minds of those who observed the ascension as to which way their Lord went when He returned to the Father in heaven. They were to observe what happened with certainty so that they would be sure that everything Jesus told them while He was in the world would come to pass as He said it would.

The resurrected Jesus Christ in His glorified body, going through locked doors and walls, hanging with His disciples, being fully visible to all of them who were with Him, speaking to them on point and being spoken to as in ordinary every day conversation, and sharing meals with them, urging the disciples to ascertain for themselves that He was real, indicates beyond reasonable doubt that getting from the earth to heaven does not take time at all with the Spirit of God involved.

The God of heavens and the earth fills the entire earth and the universe with Himself that moving something within the universe—from one extreme end of it to another extreme end of it—is faster than moving that thing from His left hand to His right hand. Since it does not take any of us much time to move things from our right hand to our left hand, it takes God even less time to do the same. Our God is certainly that big, and He accomplishes everything that fast!

That is why God is ever present in our midst in everything we do in life, because even though He is up in heaven, high and mighty, He is also here with all of us; and in everywhere in the universe in real time. God, indeed, fills the earth and the universe completely as the Bible says in *Ephesians 4:10*.

Clearly then, God's domicile—the highest heaven—is not within our physical universe; or God would have been subject to the natural laws and conditions that exist within our universe. God cannot be constrained, and as such could not exist entirely within a constrained universe—God is above everything He created, and had made everything He had created wholly accessible to Himself, giving Himself the supremacy and control over all.

God, in *Genesis 1:6*, commanded for water to be separated from water as He created space. The Bible says: *"And God said, "Let there be a vault between the waters to separate water from water." ⁷ So God made the vault and separated the water under the vault from the water above it. And it was so. ⁸ God called the vault "sky." And there was evening, and there was morning—the second day." (Genesis 1:6-8)*

God, on Day 2 of creation, spent the whole day creating space and perfecting it—in rigid framework. Through the process of creating space, God separated the water under space from the water on top of the space. He carried the water on top of space to the farthest limits of space, where it is still in existence today. This fact was beyond the comprehension of the early scientists that they dismissed it as fantasy.

They were looking for a world/universe that they could subdue and conform to fit their own imagination. Either that or they missed the story completely. God's mystery in the hands of a person of faith is clear and unobstructed; and a jewel from God. But the same mystery in the hands of a worldly person becomes another opportunity to score a personal intellectual victory; so it ends up being laced with conjectures and extrapolations, complicated theories and exotic mathematics.

Space was later, on Day four of creation, populated with celestial bodies to become our universe. The water atop the firmament (space) bounds the outer perimeters of space, causing that segment of water to effectively cease being a part of our world; and our universe.

That water is so far out of the human reach that only those who believe God and *"fully obey the LORD our God and are careful to follow all His commands—Deuteronomy 15:5"*—would ever see that water. In essence, God's abode—heaven—is outside of our universe and is a real place as Jesus Christ said it is. Jesus Christ said that He has gone back to heaven to prepare to a place for all of mankind who choose to open their eyes and accept His grace.

He also told us that if it wasn't true that heaven is a real physical place and that He was going there to prepare places for all of us, that He would not have told us so.

And in *Genesis 1:6-8*, God told us that there is a place beyond our universe which still has the water He carried out of the universe when He created space. And that the water is crystal clear because the water has no impurities in it. And here is how the Bible describes that water:

*"Then the angel showed me **the river of the water of life, as clear as crystal, flowing from the throne of God and of the Lamb ² down the middle of the great street of the city.** On each side of the river stood the tree of life, bearing twelve crops of fruit, yielding its fruit every month. And the leaves of the tree are for the healing of the nations. ³ No longer will there be any curse. The throne of God and of the Lamb will be in the city, and his servants will serve him. ⁴ They will see his face, and his name will be on their foreheads. ⁵ There will be no more night. They will not need the light of a lamp or the light of the sun, for the Lord God will give them light. And they will reign for ever and ever." (Revelation 22:1-5)*

It is just as in everything God has given to man: God retains a lot more than He gave to man. And the water atop space—that is, the water outside the universe from *Genesis 1:6-8*—is no different.

From the proportion of space to the earth, knowledge of Physics and Geometry requires that the water above space (firmament or vault) must be infinitely greater than the water surrounding the earth—*Genesis 1:6-8*—in order to have the "space" of Genesis Chapter One under vacuum; which appears to be God's intended purpose for creating space inside water in the first place.

The great volume of water above space—*that immeasurable deluge of water in the hollow of God's mighty hand (Isaiah 40:12)*—also helped to muffle the unprecedented loudness of the cataclysmic explosion of Day 4 that brought the universe to life.

The knowledge of Science and astronomy allows one to see unbelievable physical phenomena, but spiritual discernment allows one to see God at work, revealing to one what is yet to come.

All scientific discoveries are revelations of sort given to dedicated scientists who labored tirelessly in pursuit of knowledge. Therefore all scientific facts are designed to support the truth of the Bible—the truth of God.

But when any scientific information appears to contradict God and the Bible, it is not the fault of God, the revelator. It is the fault of the person to whom the revelation was given. The distortion is the person's, for whatever reason best known to the person.

Yet anyone with a discerning spirit could extract the truth which God intended out of the revelation that was distorted by any anti-Christ scientist. That is why Jesus spoke to the world in parables and spoke plainly to His disciples, declaring:

"The secret of the kingdom of God has been given to you. But to those on the outside everything is said in parables [12] so that,

"'they may be ever seeing but never perceiving,
  and ever hearing but never understanding;
otherwise they might turn and be forgiven!'"" (Mark 4:11-12)

And for anyone who would jump to the conclusion and rashly accuse Jesus Christ of not playing fair by using parables to communicate His truth to those outside His kingdom, here is God's justification for that:

God allows nations and human individuals endless time to repent and be forgiven. But anyone who continues to frolic in sin in apparent snub of the Bible, and God and His Christ, would suddenly be cut off from grace and will forever be lost. Here is God commissioning Isaiah to give that message to the God's people who had been taking God for granted:

"Then I heard the voice of the Lord saying, "Whom shall I send? And who will go for us?"

And I said, "Here am I. Send me!"

⁹ He said, "Go and tell this people:

"'Be ever hearing, but never understanding;
* be ever seeing, but never perceiving.'*
¹⁰ Make the heart of this people calloused;
* make their ears dull*
* and close their eyes.*
Otherwise they might see with their eyes,
* hear with their ears,*
* understand with their hearts,*
and turn and be healed."

¹¹ Then I said, "For how long, Lord?"

And he answered:

"Until the cities lie ruined
* and without inhabitant,*
until the houses are left deserted
* and the fields ruined and ravaged,*
¹² until the LORD has sent everyone far away
* and the land is utterly forsaken.*
¹³ And though a tenth remains in the land,
* it will again be laid waste.*
But as the terebinth and oak
* leave stumps when they are cut down,*
* so the holy seed will be the stump in the land.""* *(Isaiah 6:8-13)*

The Book of Proverbs also addresses the kind of ignorance of God and His Christ that God is trying to frustrate by this command in

Isaiah 6:9-10; which Jesus Christ echoed in *Mark 4:11-12.* Here is that passage in the Book of Proverbs:

"Whoever remains stiff-necked after many rebukes
  will suddenly be destroyed—without remedy." (Proverbs 29:1)

When confronted with the many questions of life, the scientists and astronomers see frontiers that need to be studied, deciphered and explained through physical laws for another human conquest. The spiritual man, on the other hand, sees the glory of God and immediately knows all that is to know about it, and gives praise to God.

When God talks about Him creating a new earth and a new universe, mankind—plagued by its very limited knowledge of its Maker through his lack of faith in God—do not see the possibility of that happening. But the creation of a new earth and a new universe is only a foregone conclusion. God does not say anything and not make it happen.

Our problem is that we look at everything from the natural angle. To us the universe has been in existence for billions of years and the earth was a more recent creation, because when we remove God from creation, as we have in science and astronomy, we cannot explain how the earth could have come before the universe.

God knows how we think and operate and He let us continue until we get to a point of no return in our defiance of God. Look at the following passage:

"Now then, listen, you lover of pleasure,
  lounging in your security
and saying to yourself,
  'I am, and there is none besides me.
I will never be a widow
  or suffer the loss of children.'
⁹ Both of these will overtake you

in a moment, on a single day:
loss of children and widowhood.
They will come upon you in full measure,
in spite of your many sorceries
and all your potent spells.
¹⁰ You have trusted in your wickedness
and have said, 'No one sees me.'
Your wisdom and knowledge mislead you
when you say to yourself,
'I am, and there is none besides me.'
¹¹ Disaster will come upon you,
and you will not know how to conjure it away.
A calamity will fall upon you
that you cannot ward off with a ransom;
a catastrophe you cannot foresee
will suddenly come upon you.

¹² "Keep on, then, with your magic spells
and with your many sorceries,
which you have labored at since childhood.
Perhaps you will succeed,
perhaps you will cause terror.
¹³ All the counsel you have received has only worn you out!
Let your astrologers come forward,
those stargazers who make predictions month by month,
let them save you from what is coming upon you.
¹⁴ Surely they are like stubble;
the fire will burn them up. They cannot even save themselves
from the power of the flame.
These are not coals for warmth;
this is not a fire to sit by.
¹⁵ That is all they are to you—
these you have dealt with
and labored with since childhood.
All of them go on in their error;
there is not one that can save you." *(Isaiah 47:8-15).*

The existence of the cloud in the sky, which falls back on the earth as rain, points to the existence of water beyond the universe. The separation of water from water in *Genesis 1:6* led to the water atop the firmament being pushed out of the firmament (space) before the space was populated with celestial bodies to become the universe.

Chapter 8

God's Revelation of His Mysteries to the world through the Ages

God revealed His deepest Mysteries to the men and women of faith:

God revealed His deepest mysteries to the men and women of faith, most of who had no formal education but were taught directly by God. That is why the Bible says: *"We also have the prophetic message as something completely reliable, and you will do well to pay attention to it, as to a light shining in a dark place, until the day dawns and the morning star rises in your hearts. [20] Above all, you must understand that no prophecy of Scripture came about by the prophet's own interpretation of things. [21] For prophecy never had its origin in the human will, <u>but prophets, though human, spoke from God as they were carried along by the Holy Spirit.</u>"* (2 Peter 1:19-21).

When God teaches anyone, the retention of the information learned is 100%. When God leads anyone into His mystery, the person does not only learn what God teaches them and assimilates it completely; they also remembers everything God passed on to them and could at any instance spit the same out without any inconsistences from one accounting of that knowledge to another.

The power that brought the information to the person in the first place is the same power that works the information into every fiber of the person's being, and the same power that recounts the information each and every time the need arises for the information to be recounted. And that power is the Spirit of God, working through the spirit in the person.

God's revelation does not come by way of formal education, rather it comes through the inspiration of the Holy Spirit of God, who does all the work necessary to get the information learned and retained, and precisely guides the sharing of the information learned with others, by putting the right words and descriptive details into the mouth of the bearer of the message.

However, God uses formal knowledge a person has acquired to amplify and sharpen the mystery revealed to a person so that the person can better package the information for maximum impact.

That is why Apostle Paul—who had the most formal education of any of the apostles—was brought into so many deep mysteries by the Spirit of God, including being taken to the third heaven where he *"overheard conversations that no one was permitted to tell."* That is also why He wrote extensively about the mysteries of God, some of which are of the highest intricacies, prompting Apostle Peter— the head of all the apostles, disciples and the church—to write:

"Bear in mind that our Lord's patience means salvation, <u>just as our dear brother Paul also wrote you with the wisdom that God gave him.</u> <u>[16] He writes the same way in all his letters, speaking in them of these matters. His letters contain some things that are hard to understand, which ignorant and unstable people distort, as they do the other Scriptures, to their own destruction.</u>

Therefore, dear friends, since you have been forewarned, be on your guard so that you may not be carried away by the error of the lawless and fall from your secure position. [18] But grow in the grace and knowledge of our Lord and Savior Jesus Christ. To him be glory both now and forever! Amen." (2 Peter 3:15-18)

The list of faithful messengers to the eternal word of God is endless. And throughout the ages and through each and every one of these faithful servants of God, the message is consistent. These messengers come from very different backgrounds and characters, but their mission was the same—to save humanity from its disastrous ways.

First it was Abraham, Isaac and Jacob. Joseph followed. Then Moses and Aaron and all the events that led to the exodus of the Israelites from Egypt and their forty years in the wilderness due to their disobedience.

Then Joshua took over from Moses, and led the Israelites across the Jordan into the Promised Land; and settled the various tribes as Moses instructed. The string of judges followed (Deborah, Barak, Esther, etc.).

The women of maverick faith (Rehab, Ruth, Esther, Deborah) and the prophets (Samuel, Elijah, Micaiah, Elisha, Ezra, Isaiah, Jeremiah, Ezekiel, Daniel, Hosea, Joel, Amos, Obadiah, Jonah, Micah, Nahum, Habakkuk, Zephaniah, Haggai, Zechariah, Malachi) followed.

The kings and governors (Nehemiah, David, Solomon, ..., Hezekiah, Josiah, etc.) made their contributions as well. And the apostles (Peter, Andrew, Bartholomew, James, John, Philip, Matthew, Judas, Paul, Silas, Barnabas, Mark, Jude, Titus, etc.) took over, and handed the gospel down to us.

God revealed His deep Mysteries to the Pioneers of Science and Astronomy:

God revealed some of His deep mysteries to the pioneers of science and astronomy. Everything known today in science and astronomy or in other fields of study comes through the power of the Holy Spirit of God. Human beings do not discover anything out of their own wits. Human beings are led to discoveries by the Spirit of God, through their dedication, commitment, strong desire and hard work.

The pioneers of science and astronomy were all men and women of faith, all of who were believers of God and His Christ, and were associated with the church and/or were employed by the church.

These pioneers of science and astronomy were under the church's censorship because the church was the controlling power at the time. And for that reason, the church leadership—men who were steeped in the knowledge of the word of God, and devoted to their vocation—felt the compulsion to censor whatever information was coming from these pioneers of science and astronomy.

Most of these men were severely persecuted, killed or were forced to go into exile to avoid death. And because of all these distasteful treatments from the church, some of them broke ties with the church but continued to pursue their studies and made conclusions from their observations.

The bulk of these conclusions would later form the basis for what we know today as science. And since many of these pioneers that put this body of knowledge together suffered to some degree from the church's aggression, they begin to turn their back on the church. Some among them started using their new knowledge to criticize the church and thwart the church's efforts.

At the same time these pioneers of science were struggling to stay away from the persecution of the church, civic governments were shaping up across the world with communities trying to control other communities.

The map of Europe was drawn and redrawn endless times as prominent political powers emerge to exert their dominance over their neighbors. The most aggressive ones among these vying political elements took it to the church, trying to wrestle control away from the church and subject the church to their governance.

Eventually the church capitulated and the political powers became dominant over the church's authority and influence; since everybody that belongs to the church also belongs to a country. In essence, allegiance to people's respective countries became more important to the people than their allegiance to their church.

From this point forward, it became the country first, the church second. Unfortunately, people did not realize that by placing their allegiance to their respective countries over their allegiance to the church that they were making the importance of God in their lives secondary to the importance of their political systems in their lives.

Therefore, not only have the political powers of the nations of the world succeed in bringing church authority under them, they have also succeeded in elevating their importance to their citizens over the value of faith and the church to the people.

The world's people were all jubilant that they have freed themselves from the highly restrictive commands of God, the Bible and the church. They organized themselves in free formats that allow them to decide what they want, how they want it and when they want it. They could also scuttle anything they do not like.

With these political systems roaming free and deciding for themselves how they want to manage the resources within their respective lands, it then made sense to absorb the displaced pioneers of science and astronomy and shoot for the sky, and that was exactly what they did.

But because it was every country for itself, competition for these displaced scientists and astronomers ensued among the political systems. Whoever got the most benefitted from their combined talents and went after projects that were once unimaginable.

Science led to technologies and useful products and services with improved benefits. This fruitful union between the political systems and science now has one common enemy—the church. The political systems tolerate the church because a big percentage of each their populations also has some allegiance to the church. Therefore, it is hard for the political systems to become outright antagonistic to the church and survive.

Even though it sounds like it is all gain for the people of the world because of their improved quality of life which has resulted from their dedication and love for their countries—ahead of their dedication and love for their God—it is not even close. It is all a huge minus. Humanity has reversed God's order and has begun to chase the wrong things. The same thing that brought Adam and Eve down has repeated itself once again. Human beings have once again chosen their minds over their spirits.

When Adam and Eve were following their spirit, they obeyed everything God commanded them and had a wonderful life. And they also enjoyed the constant companionship of their God. But the moment Eve decided that her mind could provide better for her, and she followed it, she fell of the cliff, pulling Adam with her. Adam was right beside her and did nothing to stop her from the path she was going.

It was the mind that ruined everything for Adam and Eve. Once again, mankind is going down that path. But this time around, mankind is heading to a more catastrophic failure—the end of time as we know time. That prophecy of God in the Bible must come to pass; because the all-knowing God foretold that future long time ago.

Today, God still reveals His Mysteries to men and women of true faith:

Today, God is still revealing His deep mysteries to the men and women of faith, some of who have the knowledge of science, astronomy and other modern fields of study. Many of these people are believers who are not satisfied with the religious status quo, and who are equally dissatisfied with the direction science and politics are leading the world in. The Bible says: My people are perishing for lack of knowledge. So the world of Christianity must diligently search through the available information and put it in proper context to advance the word of God.

The disciples of old were not devoid of knowledge. They used their knowledge of farming, fishing, and so on, to explain the things of God and His kingdom. Now, times have changed, and our professions are vast and more varied, but our faith in God is nowhere equal to theirs.

We trust now on the human systems more than we trust in God because we think we are more advanced than those believers from long ago; but we are not. Our so-called advancement is actually diminishing our faith in God. So, it is good that men and women of faith are doing something new in their professions to reaffirm the truth of God.

These men and women of faith—who have genuine interest in the truth of God, the welfare of humanity, and God's salvation for all mankind—are working tirelessly in their respective fields of specialization, bringing to the surface vital information, that would otherwise be suppressed to allow the world to continue with its anti-God agenda. They also expose the glossed over lies that has led to the exodus of otherwise good people from the church.

Chapter 9

The End must come as the Bible
Prophesied—so be ready!

The Bible is God's biggest mystery to humanity. The Bible may appear to be a simple book, but it is not. Mankind, from Adam and Eve to this generation, has not fully understood the scope and breath of the Bible. The Bible is called a living book because God created the Bible to be alive. The prints in the Bible may appear to be static prints used to capture the events and miracles of time past. But it is not so. The Bible is the word of God from the beginning to the finish, and an eternal mirror into human lives and human activities throughout human existence.

And because the Bible is a living book, it is not an ancient book. It is alive and active and God continues to make entries into it. The Bible is a dynamic book because it has life breathed into it by the creator of heaven and earth, and preserved for the benefit of God's most important creation—mankind. The Bible is the eternal record of everything God has done from the moment He decided to create the earth and the universe so they could serve God's purpose for mankind. It is the record from the beginning to the end of time.

Many Christian teachers, in reaction to the scientific facts the world uses to stifle the word of God, now prefer to say "end of this age" because they have become convinced that time will never end. That is because they do not understand scientifically how the current earth's and the universe's time started.

They believe that God started time but they lack the information that will help them firmly and scientifically establish that God undoubtedly started time. So to be safe, they configure their message to what the scientists are saying: they changed the terminology to "end of this age" and not "end of time".

Let me assure everyone, God does not make any mistakes because He is not human and not prone to mistakes. God is the absolute of everything; and everything in the world and the universe does not only point to God, they are all connected to God, in real time.

And this is scientifically speaking. And I am staking my life on it because I know with every intellectual faculty in my body that it is unequivocally true. I am connected to God in real time. You are connected to God in real time. And every life on the earth and everything else on the earth and in the universe is connected to God in real time. And this is physical connection, folks.

God started time in *Genesis 1:3* when He created light, and light forced the darkness to recede to one side of the earth because the spherical shape of the earth allowed the light to impinge on only one side of the earth on appearance. And this was by design and not out of consequence of anything God had not considered and planned for. It was all part of God's grand scheme.

And the instant of the very first appearance of light on the new earth is high noon! That is why the Bible says, *"... And the evening and the morning were the first day." (Genesis 1:5)* Because the instant of the first appearance of light on the earth is noon, naturally the evening followed, then night and finally, the morning, to complete the very first day on the planet earth. And while this very first day came and went, the earth still remained completely submerged inside the water in the hollow of God's hand.

The instant of the very first appearance of light on the new earth is the very beginning of time. Therefore, time—as we know it today, and as the world has seen it, studied it and tried to understand it throughout human existence—started with the appearance of light on the earth at the command of the Almighty God. And God, the all-knowing and the unfathomable, was sure to tell us this at the beginning of His great narrative that is the Book of Genesis.

Genesis 1:2 clearly says that the earth was completely swallowed up by water and that the surface of the water was pitch black and dark from "clouds" *(Job 38:8-9)*. And that the Spirit of God was hovering over the waters—providing the gravitational force that holds the earth in place, in space. Scientists have convinced the world that gravitational force originates from the core of large celestial bodies, relative to the size of the celestial body.

But God clearly shows us in *Genesis 1:2*— *and the Spirit of God was hovering over the waters*—that it was the Spirit of God that held the earth and the water in place, in space; and not some supposed force that the earth, by its own power, was generating to keep itself from darting around all over space. God then drove the point home in *Job 38:8-9*—

"Who shut up the sea behind doors
when it burst forth from the womb,
⁹ *when I made the clouds its garment*
and wrapped it in thick darkness,
¹⁰ *when I fixed limits for it*
and set its doors and bars in place,
¹¹ *when I said, 'This far you may come and no farther;*
here is where your proud waves halt'? *(Job 38:8-9)*

In this passage in Job, God gave us more details on how the waters of *Genesis 1:2* came to be; how He set limits for the water and contained the water and the earth that was inside it; and kept them suspended over nothing. *Genesis 1:2* says that the Spirit of God completely surrounded the water. And in the passage from Job, God in His own words says about the water of *Genesis 1:2*, "I made the clouds its garment and wrapped it in thick darkness." And this explains why there was darkness over the water in

Genesis 1:2.

1 John 1:5 says that God is light and in Him there is no darkness at all. Yet from the very first moment God created the earth, He allowed it to be covered in darkness. This constituted such a puzzle to me until God took me to *Job Chapter 38* and showed me the source of the darkness—dark clouds. But since everything God does also has symbolic meaning, I still wondered why He started out with darkness before bring in light to kick off time. Again, God gave me the answer through my spirit by taking me to other passages in the Bible, as follow:

Luke 10:18
He [Jesus] replied, "I saw Satan fall like lightning from heaven.

Genesis 3:11
"And he said, "Who told you that you were naked? Have you eaten from the tree that I commanded you not to eat from?"

John 1:4-5 *"In him was life, and that life was the light of all mankind.*
⁵ The light shines in the darkness, and the darkness has not overcome it."

Job 1:6-8
"One day the angels came to present themselves before the LORD, and Satan also came with them. ⁷ The LORD said to Satan, "Where have you come from?"
Satan answered the LORD, "From roaming throughout the earth, going back and forth on it."
⁸ Then the LORD said to Satan, "Have you considered my servant Job? There is no one on earth like him; he is blameless and upright, a man who fears God and shuns evil."

Revelation 22:5
"There will be no more night. They will not need the light of a lamp or the light of the sun, for the Lord God will give them light. And they will reign for ever and ever."

Summarizing from these passages and many others in the Bible, Satan was already on the earth at *Genesis 1:2*. And he was in the darkness. Satan was kicked out of heaven and landed on the earth where he was, and still remains, darkness and evil. And even though Satan was on the earth, God did not give him any authority over the earth or over mankind who God created the earth for.

Satan was in the world as an outcast because of his envy towards God. And while he was in the world, God used him to test the faith of all mankind *(Deuteronomy 13:3 and Job 1:8)*, and also to demonstrate God's powers to all mankind—just the same way God later used the Pharaoh of Egypt to demonstrate His might to the world. *(Exodus 9:15-16)*

When God formed the earth in Genesis Chapter One starting from verse 3, God brought in the light and chased the devil to one corner, where the devil rotated with the darkness to maintain his cover. Jesus Christ was that light of Genesis 1:3, and directly illuminated the earth from Day 1 of creation to the end of Day 3 of creation. And on day 4, the light transformed into trillions of lights and fire and sped across the open space, forming celestial clusters and different formations to form the universe that we know today.

God did not count the condition of the earth in *Genesis 1:2* as part of our world's time. *Genesis 1:2* **is "pre-time."** God counted Day One from the instance of the first appearance of light on the earth *(Genesis 1:5)*. And this starting point of time is purely scientific.

The instant of the first appearance of anything is the origination of that thing. In other words, the very nanosecond anything comes into existence is the start of that thing, and God established that for us in *Genesis 1:3*.

And because this original light instantaneously forced the darkness away from one half of the earth and continued to chase the darkness around as the earth rotated, without the darkness pushing back and regaining the coverage it lost to the light, John 1:5 says that *"the light shines in the darkness, and the darkness has not overcome it."* And there are many references throughout the Bible of this power of the light completely squashing the power of darkness, the devil and his evil.

And while on the earth, Satan deceived Adam and Eve and stole from them the dominion of the world. And from that point until now, Satan controls the entire world, getting everybody into trouble in an attempt to turn them away from God and prove his point to God—*"But now stretch out your hand and strike everything he has, and he will surely curse you to your face."* *(Job 1:11)*

Satan was on the earth right from the beginning of time—time as we know it. The darkness that covered the earth and the water all around it in *Genesis 1:2* provided a cover for Satan who lurks in the dark. Satan is afraid of light because he does not want to be exposed. And when God brought light to the earth, darkness receded to one half of the sphere that is the earth, still providing cover for Satan and his evil minions.

To demonstrate that the darkness on half of the earth at any given time is simply a convenience for the humans to take break from their work and rest up to regain energy for another day of work, God tilted the earth on its axis, making the Artic and the Antarctic circles alternate in the sharing of six straight months of light and six straight months of darkness. The nighttime is a convenience to man to get some rest and not a necessity.

The proof: People who currently live close to these circles go to sleep while there is still daylight because it is time for them to sleep and regain their energy. The scientists who live at the Antarctica during the six straight months of light have a daily routine in spite of the continuous six-month daytime.

Not having nights after the Last Day constitutes a concrete proof that the darkness that currently gives us the nighttime is not a necessity for our lives, but rather a simple convenience.

The emergence of science out of the church, coupled with the emergence of our current political systems—by seizing power from the church and relegating the church into the background—have combined to fulfil the beginning of the prophecies of the end time. And many Christians and other people of faith are oblivious that this truth of the Bible is being fulfilled.

That in itself is not surprising, because God knows that human beings always convince themselves to believe whatever makes them feel safe, regardless of how unfounded it may be. God's agenda will move on unperturbed because God achieves whatever He sets out to achieve. Here is a reminder of what is about to happen:

"Blow the trumpet in Zion;
sound the alarm on my holy hill.

Let all who live in the land tremble,
for the day of the LORD is coming.
It is close at hand—
2   a day of darkness and gloom,
*   a day of clouds and blackness.*
Like dawn spreading across the mountains
a large and mighty army comes,
such as never was in ancient times
nor ever will be in ages to come.

3 Before them fire devours,
*   behind them a flame blazes.*
Before them the land is like the garden of Eden,
*   behind them, a desert waste—*
*   nothing escapes them.*
4 They have the appearance of horses;
*   they gallop along like cavalry.*
5 With a noise like that of chariots
*   they leap over the mountaintops,*

like a crackling fire consuming stubble,
  like a mighty army drawn up for battle.

⁶ At the sight of them, nations are in anguish;
  every face turns pale.
⁷ They charge like warriors;
  they scale walls like soldiers.
They all march in line,
  not swerving from their course.
⁸ They do not jostle each other;
  each marches straight ahead.
They plunge through defenses
  without breaking ranks.
⁹ They rush upon the city;
  they run along the wall.
They climb into the houses;
  like thieves they enter through the windows.

¹⁰ Before them the earth shakes,
  the heavens tremble,
the sun and moon are darkened,
  and the stars no longer shine.
¹¹ The LORD thunders
  at the head of his army;
his forces are beyond number,
  and mighty is the army that obeys his command.
The day of the LORD is great;
  it is dreadful.
  Who can endure it?" (Joel 2:-1-11)

"¹² "Even now," declares the LORD,
  "return to me with all your heart,
  with fasting and weeping and mourning."

¹³ Rend your heart
  and not your garments.
Return to the LORD your God,
  for he is gracious and compassionate,
slow to anger and abounding in love,
  and he relents from sending calamity." (Joel 2:12-13)

"Woe to those who call evil good
  and good evil,
who put darkness for light
  and light for darkness,
who put bitter for sweet
  and sweet for bitter.

²¹ Woe to those who are wise in their own eyes
  and clever in their own sight.

²² Woe to those who are heroes at drinking wine
  and champions at mixing drinks,
²³ who acquit the guilty for a bribe,
  but deny justice to the innocent.
²⁴ Therefore, as tongues of fire lick up straw
  and as dry grass sinks down in the flames,
so their roots will decay
  and their flowers blow away like dust;
for they have rejected the law of the LORD *Almighty*
  and spurned the word of the Holy One of Israel.
²⁵ Therefore the LORD*'s anger burns against his people;*
  his hand is raised and he strikes them down.
The mountains shake,
  and the dead bodies are like refuse in the streets.

Yet for all this, his anger is not turned away,
  his hand is still upraised." (Isaiah 5:20-25)

"I watched as he opened the sixth seal. There was a great earthquake. The sun turned black like sackcloth made of goat hair, the whole moon turned blood red, ¹³ and the stars in the sky fell to earth, as figs drop from a fig tree when shaken by a strong wind. ¹⁴ The heavens receded like a scroll being rolled up, and every mountain and island was removed from its place.

¹⁵ Then the kings of the earth, the princes, the generals, the rich, the mighty, and everyone else, both slave and free, hid in caves and among the rocks of the mountains. ¹⁶ They called to the mountains and the rocks, "Fall on us and hide us from the face of him who sits on the *throne and from the wrath of the Lamb! ^{17} For the great day of their wrath has come, and who can withstand it?""* (Revelation 6:12-17)

Our Bible is not an archaic book that was written in the dark ages by unenlightened and uncivilized minds. Our Bible is pure science and the essence of science. Our Bible is the very word of the Almighty God of heaven, the earth and the universe preserved for all mankind as a road map through life and to salvation.

Our Bible is God's engineering documentation with God's blueprints for everything God ever built. And the Bible incorporates all of mankind's overt and covert activities throughout mankind's existence from Adam to the end of time. Even the things our generation is yet do, is already in the Bible. Therefore, human lives catch up to the Bible because the Bible is a foreteller of the future.

All of the world's past, present, and future activities are already captured on the pages of the Bible. So when people say that the Bible is a living Book, the Bible is not only living because it reveals things to a select few—things that the majority of the world could not see in the text.

The Bible is living because it captures all of humanity's activities as the world goes on; and even the future actions humanity has yet to take. It is all of these that make the Bible a living Book—the Book of life—the life of all human beings.

It then becomes imperative for people to desist from attacking other religions. All Christians everywhere in the world must allow the light of Jesus Christ to shine through their lives so that people of other religions would come to know that Jesus Christ belongs to all mankind. And when they discover that truth through our godly living and our complete obedience to Jesus Christ, they would not have any other place to go but to imitate our lives so they could receive the same glow they have seen in us.

What makes anyone a good Christian and a disciple of Jesus Christ is not how religious they are—that is how much of the Bible they know and how well they broadcast it. Rather, what makes anyone a good Christian and a disciple of Jesus Christ is their passion and dedication in trying to save others who do not know how to receive the same grace they have received.

Therefore, all Christians must be passionate and relentless about helping the rest of the world desire the grace of Jesus Christ that they need to be saved. That is what the apostles and the early disciples of Jesus Christ did. And their dedication and sacrifices are the reason we are saved today.

And since they did it for us, we owe it to the rest of the world, to do the same for them. That is what Christianity is all about. And that is what our great God desires of us: *"Therefore **go and make disciples of all nations**, baptizing them in the name of the Father and of the Son and of the Holy Spirit, [20] and teaching them to obey everything I have commanded you. And surely I am with you always, to the very end of the age.""* (Matthew 28:19-20)

And, Apostle Peter explains to us why God wants us to make disciples of all nations: *"The Lord is not slow in keeping his promise, as some understand slowness. Instead **he is patient with you, not wanting anyone to perish, but everyone to come to repentance**."* (2 Peter 3:9)

The mission of every Christian is not to verbally attack what others do for religion or to publicly condemn their way of life and draw a battle line. You may think that doing so is being passionate for Jesus Christ. It is not. Rather. You are working against Jesus Christ. Christ said that anyone who is not working with Him is working against Him, and anyone who is not gathering with Him, is scattering.

True Christians love all human beings and have the desire to save all. And nobody saves anyone by attacking them. When you attack people, if you do not kill them, you chase them away. But you can never bring them to desire your faith—or your ideology as the case may be. Because you have made yourself their enemy!

A disciple of Jesus Christ is a person who values all lives and has honest desire to save all of them. And anybody who truly wants to save anyone would first try to understand what the person's problems are; and come up with a technique for attempting a rescue. Trying to scare someone, who is already in danger, could easily make them let go and take a plunge into the deep end and be lost forever.

But when the person you are trying to save looks into your eyes and see genuine love and concern in them, they become willing to work with you and receive the life you are trying to help them receive. This, however, is not the same as joining them in the things they are doing, or watering down the gospel, or softening the gravity of their sin, to avoid the appearance of judging them.

When you do any of these, you are getting yourself into trouble with your maker, and there are consequences that come with that. When your love is genuine and your intent is noble, the Holy Spirit of God provides you with everything you need to be effective in your mission, and your success is guaranteed. Genuine desire for the salvation of all human beings always puts at your disposal the right tools to disciple others. It all starts with heart-felt love and genuineness.

The appearance of the supernatural light on the earth started time; and on the fourth day handed over to the natural light (subnlight). And the disappearance of the natural light (sunlight) on the earth will signal the end of time as we know it. And the Book of Revelation says that *the heavens will receded like a scroll (Revelation 6:14); and that a new heaven and a new earth will take their places (Revelation 21:17); and that there will be no more night (Revelation 22:5).*

Look at what Jesus Christ said when He was being questioned by the High Priest Caiaphas before they sent Him to His death. The Jews had brought two false witnesses who brought false testimony against Jesus so they could find the grounds to kill Him. The witnesses had testified that Jesus Christ said to tear down the mighty temple, and that in three days He will rebuild it. So Caiaphas the High priest demanded that Jesus responds to that allegation, asking Jesus whether He is the Messiah, the Son of God. And here is Christ's response:

*""You have said so," Jesus replied. "But I say to all of you: **From now on you will see the Son of Man <u>sitting</u> at the right hand of the Mighty One and <u>coming</u> on the clouds of heaven.""** (Matthew 26:64)*

Look at the passage again. Jesus Christ told the Jewish leaders that they and the entire world will see Him seated "at the right hand of God" as He comes to judge all humanity. This is an announcement to the world that the next time the world sees Him, the world will see Him seated as a king beside the King of kings—God the Father.

Therefore, when the Book of Revelation announced that God will peel back the universe with a roar, the passage is clearly saying that by removing the universe, the humans on the earth will automatically come face to face with God and His Christ who are seated on their thrones in heaven.

Removing the barrier that is the universe is tantamount to bringing God and His Christ right into the world; when in reality they have not moved an inch. God is the immovable mover. He moves everything; and never moves for anything, because He does not have to.

From His throne, He can reach anything He wants to reach within the earth, the universe and His dwelling place. God controls all of His creations through His Spirit. In addition, God has the physical reach to get to each and every one of His creations. Remember,

God has the entire universe inside of His mighty hand.

In other words, when the universe is imploded, our horizon becomes one and the same with the heavenly places beyond the universe, uniting our world with God's heavenly places, and giving us our first and continuous view of God, His Christ and the expansive and majestic God's dwelling, which would then be the only thing that completely surrounds our tiny earth.

Here is a graphic illustration of what the Book of Revelation is describing. Figure 21 shows a representation of God's dwelling place and the earth on the Last Day as prophesied in the Book of Revelation. Figure 19, which is the same as Figure 21, is the representation of God's dwelling place and the earth on Day 1 of creation in Genesis Chapter One.

Figure 20 shows the universe currently sandwiched between the earth and God's Dwelling place explaining why no one has ever seen God, because the universe was already put in place on Day 4 of creation while mankind was created on Day 6 of creation.

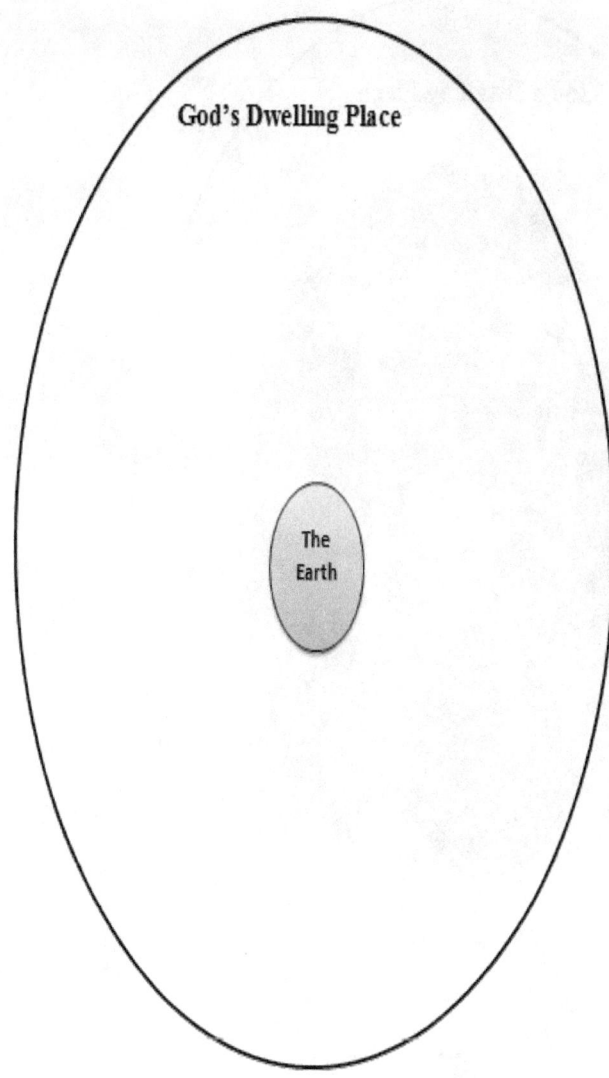

Figure 19: *Configuration at Genesis 1:2-5 (Not Drawn to Scale)*

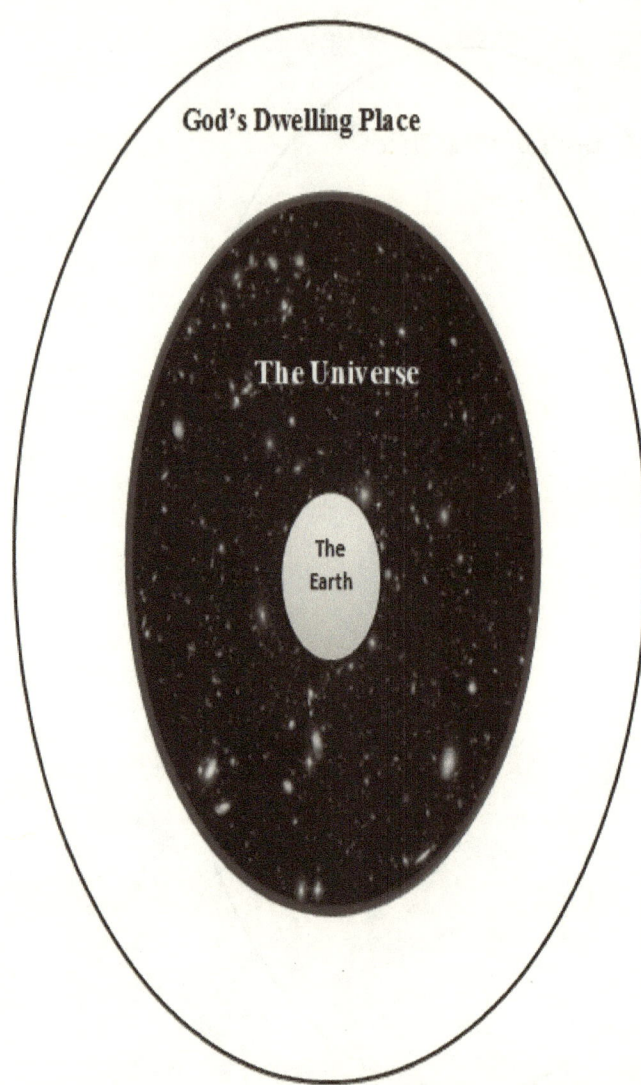

Figure 20: *Present-Day Configuration (Not Drawn to Scale)*

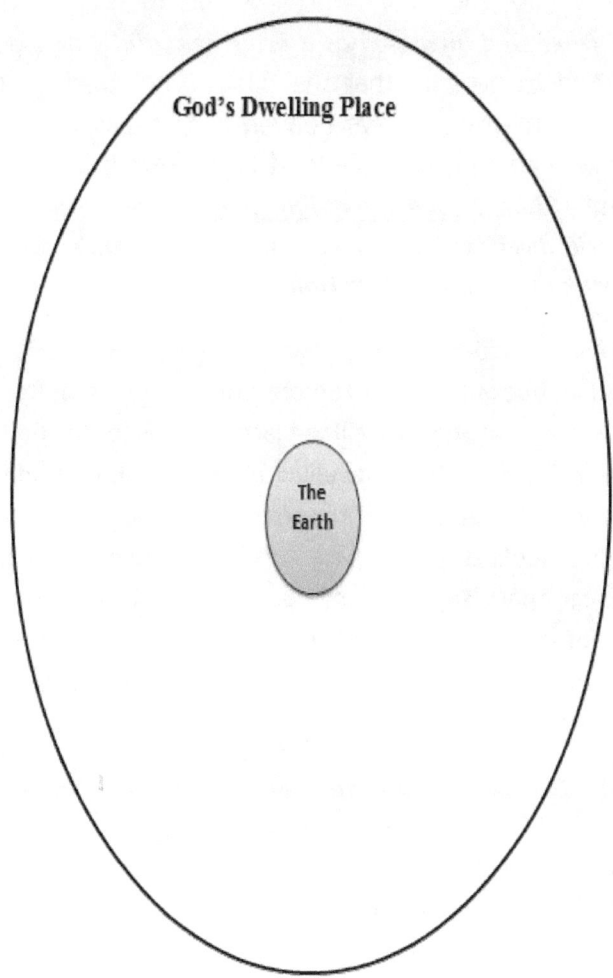

Figure 21: *End-Time Configuration (Not Drawn to Scale)*

Therefore, God will end our current world and our current universe and our current time just the same way He started them. The universe which was created last *(Genesis 1:14-19)* will be ended first *(Revelation 6:14)*. And when "the heavens recede like a scroll being rolled up", it takes with it the light of the sun, the stars and the moon, effectively ending our current time system.

Finally, "the Holy City, the New Jerusalem" will then come down *"out of heaven from God, prepared as a bride beautifully dressed for her husband."* And to be sure that the Bible is not talking about our current universe when it says heaven in this passage about the New Jerusalem coming down, a loud voice from the throne of God announced, *"Look! <u>God's dwelling place</u> is now among the people, and he will dwell with them. They will be his people, and God himself will be with them and be their God."*

According to *Genesis 1:6-8,* *"<u>God's dwelling place</u>"* is not located within our universe, but outside of it. So our universe is the buffer that is between the earth and God's dwelling place. And once this buffer is removed in *Revelation 6:14,* God's dwelling place—which currently sits atop our current universe would become earth's outer covering, physically merging God's dwelling place with ours. Currently the two dwellings are so far apart, separated by our current universe—the way the current face of the earth separates our world from hell. Here is a Scriptural evidence of the dwelling place of God currently being atop the current universe:

"And God said, "Let there be a vault between the waters to separate water from water." ⁷ So God made the vault and separated the water under the vault from the water above it. And it was so. ⁸ God called the vault "sky." And there was evening, and there was morning—the second day." (Genesis 1:6-8)

By separating *"the water under the vault from the water above it,"* God pushed the water above the "vault"—which denotes our current day space, the seemingly infinite space that we see anytime we look up into the sky. Our space appears to be infinitely expansive but it is not. And God was sure to tell us this in *Genesis 1:6-8,* only that our very limited intellect has convinced us that it is impossible for even God to push water beyond the unfathomably expansive universe of ours.

That is simply the limitedness of our minds. God did push water beyond the limits of space and He described that operation in *Genesis 1:6-8.*

That is one of His great mysteries. And the deficiency in the mental capacity to decipher this truth is not suffered by the secular world alone. That deficiency has also plagued even those who are called by the name of God due to their little faith in the awesome God who they profess.

And it is this deficiency in the intellectual capacity of the human beings to fathom the depth of the mysteries which God reveals on the pages of the Bible that had led to so many good Christians over the centuries meddling with the truth of God simply to bring God's truth to the level of the human mind, because they are in the position that allowed them to do so.

When the Bible says in *Ephesians 4:10* that Jesus Christ, who is one with God the Father, *"ascended higher than all the heavens in order to fill the whole universe"*, even the Christians and many Christian leaders think that the Bible is talking metaphorically. It is not! What the Bible says in this passage and other similar passages in the Bible—such as *Acts 17: 28*—are real and the plain and unchangeable truth of God.

This confusion may have stemmed from most human beings believing that since God is a living being, He has to be of an easily conceivable size, especially since the Bible says that human beings were made in the image and likeness of God. It is honestly difficult to understand a being in the image and likeness of a man being of a size that completely fills the earth and the entire universe, but that is who God is and more; or He would not have said that in so many places in the Bible.

Some of the questions that many would ask and readily become convinced that God cannot possibly be as large as the entire universe and have a definite form or shape are:

Can a living being have fire and such intense heat as the sun, the stars and the quasi, burning inside of Him and still be alive?

Can a living being have such diverse extreme weathers, such as extreme absolute zero temperatures as we know exist in faraway planets and parts of space; violent storms that continuously rage on some of the planets; highly toxic conditions that exist on our planets and others; and still be alive?

The Bible says that God, the source of all life, all knowledge and all wisdom, has unending capacity for everything. The Bible says that God is everything and that nothing on the earth or the universe exists outside of God. And to leave no doubt on the mind of anybody who loves the truth, God created within our world many different worlds to persuade us about his unlimited size and powers. The concept of living beings and other things living inside of another is not anything new to us. And here are some examples:

The human digestive system is a world to micro-organisms that live in it throughout a person's life on the earth. These organisms start their lives and live it out inside our digestive system. They are oblivious of the world that exists outside the person's digestive system. And within their world, they have so many different colonies and a multitude of neighbors who share whatever resources exist with the digestive system.

The food we routinely ingest and the drinks are their thunderstorms, their tsunamis and their other life's worries, depending on their nature and their attributes. They have expectations and when they are not getting it, they go crazy on us and create uncomfortable conditions for us, which may include eroding our stomach linings, giving us fever and upset stomach, and literarily evading the tissues and bones in our body in search of a better condition.

From time to time, they get visitors that are vicious and who attack mercilessly and wipe out colonies of them. That constitutes their plagues. And anytime they create discomfort in us, we chase them down with medicines and other remedies. And they go on the run to save themselves. Our medicines and therapies constitute their mortal dangers and reprisal for daring to make us sick or uncomfortable.

Even some of the microorganisms that live inside of our digestive system have other microorganisms much smaller than them living inside of them and colonizing their bodies. The T-4 Bacteriophage is one such microorganism. The T-4 is a virus that affects bacteria, and they affect E-Coli which live inside the human digestive system.

Science claims that between 200 and 300 T-4 bacteriophages can fit into a single E-Coli bacterium. And since viruses only come alive inside a compatible biological host, the body of E-coli becomes a world for the T-4 virus, thereby creating potentially millions of worlds within the world of the E-Coli in our digestive system in the event of T-4 infection of the E-coli in a person's digestive system. God puts these signs in our world to help us discover the truth in our Bible.

However, our universe physically sits inside God's mighty hand— and not inside God's holy body. But our universe is completely filled and saturated with the Spirit of God. In just the same way the earth and our solar system sit and rotate in one swirling arm of the milky-way, our universe sits on one hand of God and rotates before God. Therefore, when we trust the Bible and believe that our God is literarily much bigger than our entire universe, we then easily see the truthfulness of all the other claims in the Bible, such as the following:

Isaiah 40:22
"He sits enthroned above the circle of the earth, and its people are like grasshoppers. He stretches out the heavens like a canopy, and spreads them out like a tent to live in."

Matthew 5:35
"Again, you have heard that it was said to the people long ago, 'Do not break your oath, but fulfill to the Lord the vows you have made.' [34] But I tell you, do not swear an oath at all: either <u>by heaven, for it is God's</u>

*throne; ³⁵ or by **the earth, for it is his footstool**; or by Jerusalem, for it is the city of the Great King.*

Contrary to popular belief, God will not give the human beings the benefit of bringing an end to human existence on the earth, through say, a nuclear war or other destructive human escapades. It will not be the act of human beings that will bring our current age to an abrupt end. God would not allow the humans that power. While a nuclear war could end all human lives on earth, it would not affect the sun and all the celestial bodies in the universe in any way whatsoever.

The end of time is bigger than the collective human will; and our mess. God started time in *Genesis 1:3-5* and God will end time as prophesied in various places in the Bible. And that end will come suddenly and swiftly as God promised in the Bible. And it will come when the world is confident in itself and least expects that kind of surprise. Moments before the time the end comes, the world will be confident it could divert any asteroid on collision course with the earth.

At the very time the end comes, nations of the world would be confident they have everything under control. Human elements would only add to the confusion but would not contribute in any way, shape, or form to the destruction God has destined to end it all. This is why the Bible says that it will be like the great flood of Noah's time; and that the leaders of the nations and everybody else alike would be running to a hiding place and would still fell unprotected.

That is why Jesus Christ Himself reminded humanity of the flood of Noah and Lot's wife. The chaos of the tribulations would be going on through human conflicts as the Bible predicted, but none of those would bring the end. The end would come of God's determination and God's timing, and God's origination. And the end would have nothing to do with anything the humans could do.

God single-handedly started the earth, the time, space and the universe. And God will also single-handedly bring it all to a screeching halt as prophesied in the Bible.

Everything we now see and enjoy and take for granted was God's to give. They will all be taken away through the direct acts of God and would not be aided in any way, form or shape by the act of man. The only place the acts of men have in the end time is in creating confusion and destabilizing the world systems, but not ending them.

God has enough weapons pointing at our world to accomplish His plan for Himself. It was all Him in the beginning. And it will be all Him in the end. And for those who believe that they will use the infrastructures they have put in place to either stop God or aid God, none of it will work.

God will immobilize all human contraptions and turn them into trash when He decides to reel it all in. Just as it was in the time of Noah, it will be at the time of the Last Day. Not a single man-made object that was not already in the ark survived the great flood.

Everything that was man-made was buried deep under the earth when it was all over. Even the ark that saved Noah, his family and all the rescued animals got hung-up on top of Mt Ararat. And Noah and family had to start life afresh with no man-made contraptions from before the flood.

Therefore nothing of value would make it out of the Last Day's destruction. Those who survive that Day will inherit a new earth— our earth which is now transformed by the New Jerusalem. Our universe will implode and vanish forever, but our earth would be transformed and continues into eternity with God and His Christ. And here is why the earth continues:

"The LORD reigns, he is robed in majesty;
the LORD is robed in majesty and armed with strength.
The world is firmly established; it cannot be moved.
Your throne was established long ago;
you are from all eternity." (Psalm 93:1-2)

And here is a sample of what awaits mankind and its world:

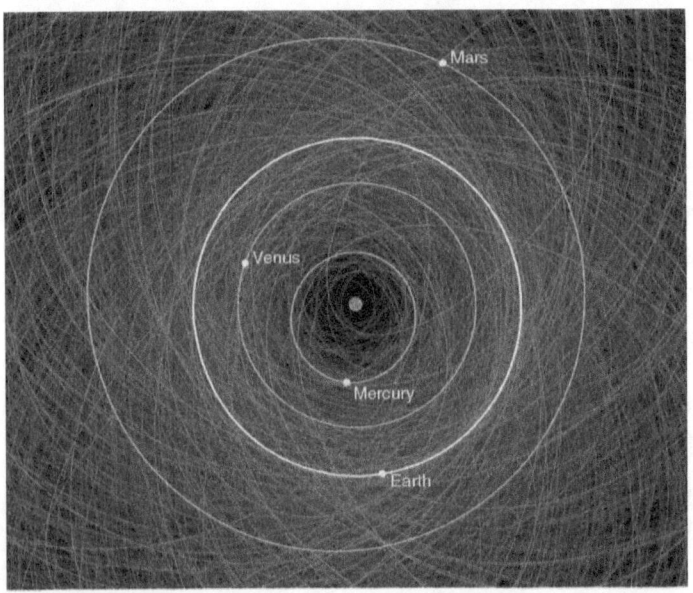

Figure 22: potential-hazardous-asteroids-crop

The picture above is an artistic illustration of the asteroids floating around our solar system with the potential to smash into the earth and obliterate life as we know it. Just as God declared in the Bible that hails are reserved for the day of war, the asteroids are also reserved for the day of destruction of the earth. God can change the orbital patterns of any number of these loose objects at any time, and orient their trajectories to coincide with that of our earth for a fateful collision—or a number of them. God always achieves His purpose without failing, and no human scheming can get in the way of that.

And here is the kind of result a substantial asteroid collision with the earth could produce:

Figure 23: The Tunguska explosion flattened some 500,000 acres of Siberian forest on June 30, 1908. This image is from the Leonid Kulik expedition in 1927

Science has projected that it was a similar event that caused the extinction of the dinosaurs. However, the Bible says that the dinosaurs coexisted with mankind—the ancients. Their extinction therefore took place while mankind was already on the planet earth. Whatever caused their extinction did not result in the extinction of mankind, as well. Therefore, that claim of science is also questionable! Here is that passage about the dinosaurs coexisting with the human beings from the Bible:

*""Look at **Behemoth**,*
*   **which I made along with you***
*   and which feeds on grass like an ox.*

[16] What <u>strength it has in its loins</u>,
  what power in the muscles of its belly!
[17] Its <u>tail sways like a cedar</u>;
  the sinews of its thighs are close-knit.
[18] Its <u>bones are tubes of bronze</u>,
  its <u>limbs like rods of iron</u>.
[19] <u>It ranks first among the works of God</u>,
  yet its Maker can approach it with his sword." *(Job 40:15-19)*

Most of science's guess works are mere science-fiction. Yet they are passed on to the world as scientific facts. We must separate true science from science-fiction or more of God's people would be deceived by the deceptions that are currently being passed around as truth. Science is truthful knowledge that supports all of God's declarations and proclamations in the Bible; and science-fiction, passed on as science, is only a misguided fallacy and must be brought to light to shame its pretense and strip it of all authenticity. The word of God will prevail on its own merit. God is greater than everything He had created because God is infinity plus everything He had created, thus

$$E_{God} = E^{\infty} + \sum_{n=1}^{\infty}(E_n)$$

Where,

E_{God} = God's energy

E^{∞} = infinite energy

E_n = Total energy of its kind available on earth and in the universe, visible and invisible

n = The number of different energy types available on the earth and in the universe

This equation is the very first law of thermodynamics and must be noted as such!

Love and Peace!

ABOUT THE AUTHOR

My life is a laboratory. And all human beings are designed as such by the all-knowing God. The only difference among us is that while some willingly become part of life's experiments, some view it from the sidelines.

The best lessons we each learn in life comes to us directly and not through a teacher in an academic setting. We all learn and mature in our experiences by trial and error, just like a scientist in the laboratory. But we are not only the scientist, we are also the test specimen and the laboratory facility & instrumentation—all rolled into one.

And when we are in tune with our spirits, it becomes more verification than 'trial and error' because through our spirits, God feeds us great knowledge about our lives, the things around us and deeper mysteries than we ever thought possible.

Most of my books happened that way. Information came into my mind and takes residence. I soon become aware of it and try to know more about it. As I explore it, it deepens and more is downloaded onto my spirit. And intuitively, I am led to its verification. Once verified, it becomes common knowledge to me.

God has been unbelievably good to me by opening windows to me into great mysterious, such as I have been writing about in my many books. There is hardly a day that I am not writing books. I work on several titles simultaneously, capturing the information as soon as it enters my mind.

Ifeanyi Chukwujama

Other Titles form this Author:

- Who is God!
- What is Love!
- Christ is in Everyone!
- Christianity is Life; not a Religion!
- The Singleness of God!
- Overcoming Your Trials!
- Live the Abundant Life!
- Science, Evolution and God!
- Reflections of Life!
- The Rapture, the Tribulations and the Church!
- The Big Bang: and Jesus Christ birthed the Universe!
- Government is a spirit and the Beast; Science is the False Prophet!
- Scientific Proof that the Earth & Water Existed before Time, Space & the Big Bang!
- The God of Science
- Whoever says that Sex is Good is a Liar!
- Revive Your Marriage instantly with God